Neural Networks

Easy Guide to Artificial Neural Networks

(Artificial Intelligence and Neural Network Concepts Explained in Simple Terms)

Laurie Thomas

Published By **Tyson Maxwell**

Laurie Thomas

Neural Networks: Easy Guide to Artificial Neural Networks (Artificial Intelligence and Neural Network Concepts Explained in Simple Terms)

ISBN   978-1-7752672-7-0

No part of this guidebook shall be reproduced in any form without permission in writing from the publisher except in the case of brief quotations embodied in critical articles or reviews.

Legal & Disclaimer

Table Of Contents

Chapter 1: Introduction To Neural Networks

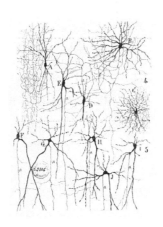

Why discover approximately neural networks?

In this segment, we are capable to discuss the development of synthetic neural networks (ANNs), which have been based completely mostly on the functioning of the human thoughts.

Over the years, many researchers completed experiments so that they may apprehend human intelligence. This research is now very beneficial in computing. In this economic catastrophe, we're capable of provide an explanation for how scientists use natural neural networks to create synthetic neural networks. We ought to have a have a look at the ones natural neural networks which can be the bases of artificial neural networks

As you recognize, it is very easy to use neural networks in computing. They were advanced near a complicated model of the human apprehensive device. You can don't forget them as simplified models of our brain capabilities.

From Neurological Research to Artificial Neural Networks

Scientists have commonly been interested in the complexity of the human thoughts. After a few years of big studies, we are in the end able to witness the improvement of this studies. Before the present day-day generation, we truly could not understand how the human thoughts capabilities. The first ANNs have been evolved in the Nineteen Nineties.

At this time, lots much less information approximately thoughts talents, together with perception and the manner intelligence works, modified into to be had. Research focusing on unique sicknesses and accidents has led us to apprehend how our brains control motion. Our information of those particular responsibilities, as completed by means of using the factors of our brains, changed into very confined.

Research allows us understand of ways the ones thoughts additives help us manipulate movement and exceptional crucial features. We now recognize which components of our brains are associated with specific kinds of accidents.

We conceived of the character hemispheres as being nicely-defined, specialised systems.

You can see this inside the following determine, which explains the localization of severa mind capabilities.

Also, within the following parent, you may apprehend the features of each element of the thoughts.

However, there can be moreover the issue of the changing of electrodes inner actual-existence information collection. Although animal experimentation is viable, killing an animal within the call of technology is instead arguable, even though it is finished for a noble motive.

It's additionally now not possible to draw an real end regarding the human mind primarily based on the consequences of animal locating out. There are versions among human and animal brains, in particular within the musculoskeletal and circulatory structures.

Let's check the only of a kind strategies that the creators of neural networks have used to paintings with the most nice abilties and residences on the mind, as they have got developed through evolution. Our brains consist of neurons that paintings as separate processing cells.

One neuroscientist defined the human mind as a community of associated factors. They generate and send manage indicators to every a part of the body. Indeed you will have a look at more about the systems of natural neurons and their artificial equivalents, which can be the center components of the shape of a neural network.

Observe, within the following decide, how an man or woman neuron is remoted from the internet of neurons that constitutes the cerebral cortex.

We will talk the elements demonstrated inside the above parent in statistics later.

As you could see, neural neurons have an complex and diverse manufacturing. Artificial neural networks have trimmed down this form. And they're furthermore much less complex in positive areas, which encompass the area of interest. Even despite the fact that right right here are many versions, you could use synthetic neural networks to duplicate complex behaviors.

As you may see inside the final figure, the neuron can be reproduced graphically so that we're able to see a real neural mobile. Which can belong to the thoughts of a rat or a human, because they're quite similar. Modeling a clean ANN with a clean digital device may be very possible.

It's very easy to version both features as one laptop algorithm that show us their sports sports. The first neural network became built as an virtual machine, known as a perceptron.

Let's communicate approximately how natural facts is used in the hassle of neurocybernetics to boom neural networks. The scientists that created the primary neural networks understood the actions of the natural neurons. The maximum essential aspect they located c modified into approximately the approach with the resource of using which one neuron passes a sign to any other neuron.

The scientists said that, while processing records, the ones big and complex cells that contend with communication the various neurons are the most crucial. The synapses also are essential contributors in the system. They're very small, so optical microscopes are required to see them.

A British neuroscientist proved that, at the identical time as the neural signal goes via the synapse, chemical materials referred to as neuromodulators are engaged. They are constantly launched at the final component of the axon, from the neurons that switch the records, that's then sent to the postsynaptic membrane of the receiver (a few one of a kind neuron).

Teaching any neuron typically relies upon at the power of a sign that is sent through manner of manner of an axon from a transmitting cell, so that a totally large or very small quantity of the mediator is sent to the synapse in order to get keep of that sign.

Questions

1. Describe the biological shape of a neuron?.

2. What is the number one feature of a neuron.

3. List the features of every element of the brain.

4. What are the abilties of temporal lobe.

Chapter 2: Structures Of Neural Networks

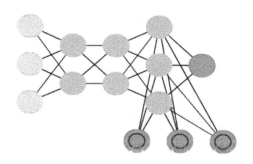

Building a Neural Network

When we have been younger, we attempted to recognize how the area capabilities: for instance with the resource of breaking an alarm clock into quantities, or taking off a tape recorder to appearance what lies inner. In that vein, permit us to strive provide a smooth rationalization of methods a neural community is constructed. As we defined in the previous section, a neural network is a gadget that plays calculations, based totally on the sports of factors called neurons.

Artificial neural networks are generally primarily based on many lot of neurons. On the alternative hand, natural neurons are associated in a much less complex manner. The ANN model of a few issue like a actual fearful machine can be very tough to govern.

In the subsequent parent, you could see how an artificial neural network is based totally on different shape and schema of our real -life involved structures.

You might in all likelihood examine that the structure isn't very smooth, however but it in reality's very complicated. It's like a huge wooded area.

Artificial neural networks are always constructed without a doubt so their structures can be clean to hint, and moreover in order that they might be used and produced very economically. In reality, the ones systems may be flat (one- or -dimensional), and often repeated, with layers of artificial neurons. They have a properly-described motive, and are related constant with a completely clean form. As you can see inside the following discern, which indicates a common neural community shape, an ANN is easy in assessment to a organic community. There are three factors which have an effect at the homes and possibilities of an synthetic neural network:

1. The elements which may be used to gather the community (how the neurons look and art work);

2. How we can connect the neurons with every other; and

three. How to establish the arguments and the parameters of your community through the gaining knowledge of method,

The Construction of Artificial Neurons

The easy homes substances used to create neural community are synthetic neurons, and we ought to take a look at them in high-quality detail. In the final segment, we discussed natural neurons.

In the subsequent determine, you could find a smooth depiction of a neuron. Not each neuron looks like this, but maximum do.

In the subsequent discern, you could see a organic neuron that is a component a rat's cerebral cortex.

It's very tough to examine the versions among an axon, that could offer signs from a particular neuron to all of the other neurons, and a dendrite, which serves a unique cause from the maze of fibers visible within the determine. Artificial neural networks have all the competencies critical for completing responsibilities they're characterised thru many inputs, however one output only. The symptoms of the enter, Xi, I = 1 or 2 or three

...n, and the sign of the output, y, might also moreover additionally take incredible one numerical price. In reality, the obligations might be solved through statistics that is the end result or the output of a specific protocol. Generally, every input and output are associated with a particular definition for any given sign. In addition, the sign extension is used, genuinely so the selected values of a signal inside a network will no longer be out of doors of an agreed-upon protocol or variety; as an instance, from 0 to 3, and so forth.

Artificial neurons perform committed sports sports on enter indicators and bring output symptoms (really one for a single neuron). This method forwarding them to the possibility neurons or to the output of the network. This is referred to as community challenge; used to reduce the functioning to its easy neural factors. This is based completely totally on the reality that it transforms an input, x, of information into an output, y, of facts through making use of the recommendations which can be placed and additionally assigned on the time the network is created. You can see how this works in the following decide.

They can discover ways to use the ones coefficients that are synaptic. The neurons mirror the complicated biochemical and bioelectric techniques that take location in organic neural synapses. These kind of weights constitute the idea of a training community, which can be changed.

Adding variable weight coefficients to a neuron form makes it a learnable unit. You can take into account the ANNs as pc processors with dedicated abilities, as described below. Every neuron receives many enter indicators, X_i, and, on the basis of the inputs, determines its solution, y, with a unmarried output sign. A weight parameter known as WI is hooked up to break up neuron inputs. It expresses the diploma of significance of information that involves a neuron the usage of a selected input, x_i. A sign arriving thru a particular input is first modified with the usage of the load of the enter.

Most frequently, the alternate is primarily based completely on the fact that a signal is certainly extended via the burden of a given enter. Consequently, in later calculations, this sign will take part in some distinctive shape.

The sign could be very strong if the weight price is better than 1 or smaller than 1. This sign from a selected input will appear inside the contrary indicators from the inputs if the load of the enter has a price of lots much less than 1. Inputs with horrible weights are described thru neural network customers as inhibitory inputs; humans with superb weights are known as excitatory inputs. Input signs and symptoms are aggregated in a neuron.

Networks use many techniques of aggregating input indicators. In reality, aggregations are actually including input indicators to decide inner signals.

This is referred as cumulative neuron stimulation or postsynaptic stimulation. This sign may be moreover defined as a network charge. Maybe a neuron offers a further aspect, impartial of enter signs, to the created sum of alerts.

We can name it a bias, and it truly goes through the learning techniques. So, a bias could be taken into consideration as a in addition weight associated with inputs, and it gives an inner signal of consistent fee identical to at the least one.

A bias permits inside the formation of a neuron's homes at a few level within the studying segment, while the feature houses do now not want to skip through the start of the coordinate gadget.

The following parent depicts a neuron with a bias. The sum of internal indicators prolonged by the usage of weights, further to a bias, can also every now and then be sent right away to its axon and dealt with as a neuron's output signal. This works nicely for linear systems,

a few thing like adaptive linear networks. In a community with greater abilties, together with a multilayer perceptron, the output of the neuron sign can be calculated thru many

talents. The picture ʄ() or ɸ() is used to symbolize the feature . Below, the determine depıcts a neuron, such as both input signal aggregation and output sign generation. Function ɸ() is a characteristic of a neuron.

Other traits of neurons exist as properly. But a number of them are selected, so that the conduct of an synthetic neuron resembles that of a organic neuron, however with tendencies that still can be decided on in a way that guarantees the overall overall performance of computations carried on via using a neural community.

At all instances, function ɸ () constructs an critical element among a joint stImulation of a neuron and its output signal. Knowledge of the inputs, weight coefficients, enter aggregation strategies, and neuron developments, allowed to unequivocally define the output sign at any time, commonly assumes

that the way seems proper now, contrary to what takes location with herbal neurons. This will help the ANNs replicate adjustments within the enter signals at once on the output. This is a in reality theoretical assumption. After enter signals alternate, even in virtual realizations, a while is wanted to set up an appropriate charge of an output sign with an correct sufficient covered circuit.

More time may be critical to obtain the same effect in a simulation: a computer imitating community sports desires to calculate all values of all symptoms on all outputs of all neurons in a community.

This could probably require some of time, notwithstanding very rapid computers. You'll no longer be aware about neuron response time in discussions of network functioning as it's a trivial hassle in this context.

This neuron supplied in this discern is the usual model with a view to be used to create a neural community. In reality, this neural network cloth is made from neurons known as perceptrons. This neuron is said by means of the use of manner of the aggregation characteristic which encompass smooth summing of enter signs, advanced thru using weights, and uses a nonlinear switch characteristic with a wonderful sigmoid shape.

Radial neurons are once in a while used for special functions. They contain an regular approach of enter facts aggregation, use specific houses, and are taught in an uncommon way.

We will now not pass into complicated information approximately these unique neurons, which can be used specially to create unique networks called radial foundation abilties.

The Biological Neurons Model

After some years, scientist Schutter attempted to version in the form and functioning of genuinely one neuron, the Purkinje, in detail. The version uses an electric system that, in step with Hodgkin and Huxley, modeled bioelectrical sports activities of character fibers and the cell membranes of neuron soma.

After considering particular research on the functioning of so-referred to as ions channels, he have become a hit in producing the shape of a real Purkinje cell with superb accuracy. The model became out to be very complex and worried high priced calculations.

For instance, it required 1,six hundred so-known as cubicles (cell fragments treated as homogeneous components containing precise materials in particular concentrations), 8,021 fashions of ions channels, more than nine one-of-a-type complicated mathematical descriptions of ion channels relying on voltage, more than 30,000 differential equations, greater than 19,000 parameters to estimate the tuning of the model, and a unique description of cell morphology, primarily based on unique microscopic pics. It's no wonder that many hours of continuous work on a huge supercomputer have been needed to simulate numerous seconds of "existence" of this sort of nerve mobile.

Despite this problem, the effects of the modeling have been very lovely, and unambiguous. This attempt at faithfully modeling the form and motion of a real herbal neuron turn out to be a success, but honestly too steeply-priced for growing realistic neural networks for giant use.

When researchers used best simplified fashions. Despite their, we realise that neural networks can remedy tremendous problems efficiently, and they even allow us to attract interesting conclusions about the behavior of the human mind.

How They Work

The first description of ANNs shows that each neuron possesses a specific internal memory (we are capable of represent them by means of the usage of the values of modern-day weights and bias) and certain abilities to convert input signs and symptoms into output indicators.

It bears noting that the ones abilties are as an opportunity limited. A neuron is form of a reasonably-priced processor inner a gadget that consists of probably loads of such factors.

ANNs are beneficial additives of structures that can technique very complex information-based completely responsibilities. A neural network is the give up quit end result of the constrained quantity of statistics amassed by using a single neuron and its bad computing talents. It consists of severa neurons that would act most effective as an entire.

Thus, all of the abilties and houses of neural networks stated in advance result from collective performances of many associated elements that represent the whole community. This uniqueness of pc era is called massive parallel processing.

Now, permit's have a study the operational data of a neural community. It's smooth from the above speak that the network software software, the statistics that constitutes expertise database, and the information may be calculated, and the calculation strategies are all virtually allocated.

It's not viable to issue to a place in which particular statistics is stored, notwithstanding the reality that neural networks may additionally furthermore feature as reminiscences, specifically as so-called associative reminiscences have tested incredible general overall performance. It's additionally not possible to attach certain areas of a network to a given part of the set of regulations that modified into used : for instance, to suggest which community factors are liable for preliminary processing and assessment and which elements produce final network outcomes.

We will now observe how a neural community works and what roles the single factors play inside the entire operation. You may additionally expect that all community weights are already decided (for instance, that e education method has been accomplished).

The critical way of education a community is as an opportunity complicated procedure. We begin our evaluation via manner of the from the point wherein a brand new task is supplied to a community. The challenge may be represented through some of enter signals appearing the least bit inputs.

The signals may be represented by manner of using red dots. These enter signs collect the neurons within the input layer. These neurons typically do no longer technique the signals; they best distribute them to the neurons in the hidden layer. The super nature of the enter layer neurons that satisfactory distribute signals, in desire to approach them, is commonly represented graphically via using using numerous sorts of symbols (e.G., a triangle in area of a square).

The next step entails activation of the neurons in the hidden layer. The neurons use their weights (because of this using the information they consist of), first to alter the enter indicators and mixture them, and then, consequently to their trends calculate the output signals which can be directed to the neurons inside the output layer.

This degree of statistics processing is crucial for neural networks. Although the layer may now not upward thrust up externally (the indicators will not be registered on the enter or output ports), this is the layer in which most of the challenge-fixing hobby is completed. Most of the network connections and their weights are positioned a number of the input and the hidden layers.

We can say that the most of the statistics in the schooling method is on this layer. These symptoms could be supplied through the layer (hidden). Neurons don't have direct contradictions, in evaluation to the enter or the output signals—each sign could have a meaning for the project this is solved—however with this method, the layer of the neurons offer not whole merchandise.

That is, signs specify the task on this form of way that it's far extraordinarily clean to use each of them.

By jogging with the performance of the network at the last stage of solved challenge, you can see that the layer of neurons will take abilities in their talents to sum up the signals and their homes to assemble the closing answer at the network output ports.

In exclusive words, a network constantly works as an entire, and all its factors make a contribution to performing all the duties of the network. This device is just like a holographic reproduction, in which you could reproduce a complete photograph of a photographed object using the quantities of a broken photographic plate.

One of the advantages of network standard overall performance is its terrific ability to artwork well, even after a exquisite part of its factors fail. One scientist has taken his a number of his network's abilities (just like the letter recognition approach) after which examined them as he damaged an increasing number of in their elements. These networks have been particular digital circuits. Though Rosenblatt may damage an important part of a community, it might preserve to characteristic nicely.

The failure of a higher extensive form of neurons and connections need to cause the tremendous of standard standard overall performance to visit pot, in that the broken a part of the community may also need to make extra errors (for example, to spotting O as D) but it'd now not fail walking.

Compare this behavior to the truth that the failure of a unmarried detail of a modern-day digital device, collectively with a computer or tv, can reason it to prevent strolling truly. More than plenty of neurons in the thoughts die every day for hundreds motives, however our brains hold to art work unfailingly inside the direction of our lives.

The Capabilities of Neural Network Structure

You can recall the relationship the numerous shape of a neural network and the responsibilities that it is able to carry out. We additionally comprehend that the neurons described in advance than are used to create neural networks.

Network structures are created with the resource of manner of connecting outputs of high quality neurons with inputs of different neurons, based mostly on a particular layout. The stop end result is a tool of parallel and max concurrent processing of numerous data. When preserving some of these factors, we normally choose out layer-primarily based totally networks, and connections among layers are made on a one-to-one basis. Obviously, the particular topology of a network (the type of neurons in layers) have to be based at the styles of duties the network will method.

In this idea, the rule is straightforward: the greater complex the mission, the more neurons are had to remedy it. A network with more neurons is clearly greater smart. To be realistic, this concept isn't always as unequivocal as it appears. The concept of neural networks includes big works, proving that choices concerning the network shape have an effect on its conduct a ways much less than predicted.

This paradoxical declaration derives from the truth that behavior of a community is real basically via the network education approach, no longer by way of manner of its shape or the variety of things it contains.

This explains how a properly-taught neural community, which has a particular shape, can remedy obligations in a greater green manner than a badly skilled community with a right form. Many experiments done on neural network systems, created by means of manner of way of randomly determining which elements be part of, and in what manner. Despite their casual designs, the networks had been capable of fixing complicated obligations.

Let's take a higher have a observe the essential effects of this random layout. If a randomly designed network can attain correct consequences no matter its form, its education manner can for that reason allow it to adjust its parameters to function as required, based totally totally on a designated set of guidelines. This technique that the device will run efficaciously, notwithstanding its completely randomized shape.

These experiments were first completed inside the early Nineteen Seventies. The scientist flipped the cube or drew straws and, based totally on the final consequences, associated positive elements of a network together. The ensuing shape emerge as without a doubt chaotic. After schooling, the network need to treatment duties efficaciously.

The scientist's opinions of his experiments had been so top notch that other scientists did now not be given as genuine alongside with his results have been viable until the experiments have been repeated. This device, which end up just like the perceptron built by means of way of Rosenblatt, have grow to be advanced and studied spherical the world.

Networks that do not have a massive connection can continually learn how to remedy duties efficiently; however, the training method for a random network is extra complicated and time-eating at the identical time compared to the teaching of a network whose structure within reason associated with the venture handy.

It's exciting to be aware that philosophers were also interested by the researchers' effects. They claimed that the very last outcomes proved a concept given, and then later modified, thru extraordinary scientists.

The researcher proved that this idea is technically feasible, essentially, within the shape of neural networks. Another hassle is whether or no longer or not the idea works for all people.

The other scientist claimed that early abilities amounted to not whatever, and newly received understanding became everything. You can't touch upon the researcher's assertion, however we do apprehend that neural networks gain all their facts most effective through manner of getting their gaining knowledge of techniques adjusted to the undertaking form.

Of direction, the network form need to be complicated sufficient to allow "crystallization" of the desired connections and systems. A community this is too small will never study some issue due to the truth its "intellectual capability" is insufficient. This critical trouble is the extensive range of things involved, now not the layout of the form.

For example, no individual teaches relativity precept to a rat even though the rat may also moreover were skilled to find out a manner thru complicated labyrinths. Similarly, no human is programmed at starting to be a doctor, an architect, or a laborer. Jobs and careers are alternatives. No declaration about equality can change the truth that some humans have awesome intellectual property and some do not.

You can follow this idea to community layout. You cannot create neural networks the usage of early era; however, it's miles now not too difficult to create a cybernetic troglodyte, which has so few neurons that it cannot analyze some thing. A network can perform extensively numerous and complicated duties if it is huge enough. Although it seems that truely a network can't be too huge, a larger length can create complications.

Studies show that a community will remedy a problem; a neural form is truly more precious. A fairly designed form, which suits the trouble' necessities at the start, can shorten mastering time appreciably, and decorate consequences.

This is why we need to talk about the improvement of neural networks, however the fact that it can no longer offer the solution for all varieties of manufacturing problems. Choosing a method to a manufacturing trouble without sufficient information is as an alternative hard, if no longer not possible.

Constructing neural networks simply so it may be tailor-made to any shape is much like the trouble of the green software program application software engineer who's forced with the aid of the device message mentioning "press any key" . What key? You also can snicker approximately that, however we pay interest a comparable question from our graduate college college students: what is "any structure" of a neural community?

We have to now be conscious some facts about commonplace neural community systems. Not all elements of all structures are sincerely understood. You can begin with the useful resource of categorizing normally used community structures to two sub-training: neural networks with and with out feedback.

Neural networks with out comments are frequently referred to as feed-in advance types. Networks wherein signals can flow into for an endless quantity of time are known as recurrent. These indicators start going from the enter and the statistics relevant to this trouble will arrive within the neural network, to get the output wherein the neural network will offer a cease end result. These kinds of networks are the most often used.

Recurrent networks are characterized through feedback. Signals can flow into amongst neurons for a totally long term in advance than they obtain a set usa. In oftentimes, the network can't offer constant states.

The connections supplied as crimson (outside) arrows are feedbacks, so the network depicted is recurrent. Recurrent community homes and competencies are greater complicated than the ones of feed-ahead networks. Additionally, their computational potentials are astonishingly great from those of different kinds of neural networks. For example, they are able to remedy optimization issues.

They can look for the brilliant possible solutions—a assignment that is almost not viable for feed-ahead networks.. In Hopfield networks, the handiest and great kind of connection among neurons is remarks. In the past, the development of the solution to the famous journeying salesman trouble with the aid of using a Hopfield community became a international sensation. Given a hard and fast of towns and the distance among every viable pair, the visiting salesman problem includes locating the nice viable way of travelling all of the towns precisely as fast as in advance than returning to the vicinity to start.

A preference to this trouble using neural networks become provided for the number one time in a paper via Hopfield and Tank.

Questions

1. Describe the distinction among natural and artificial neural networks.

2. What are the talents of neural community shape?

Chapter 3: Teaching Your Networks

Neural network pastime may be classlfled into severa levels of mastering, for the duration of which the community collects the statistics had to decide what it will do and how, and the steps of everyday artwork at the same time as the community need to resolve dedicated new obligations depending at the amassed information.

The most crucial issue to knowledge how a community works and the abilties it has is the mastering device. Two variations of reading may be superb: one that calls for a trainer and one that does not. We are going to speak about mastering with out a teacher inside the subsequent financial disaster.

This financial catastrophe will popularity on gaining knowledge of with a teach. Such gaining knowledge of is based totally totally on giving the community examples of accurate actions that it must learn how to mimic. An example normally includes a selected set of input and output indicators, given via a instructor, to expose the predicted response of the community for a given setup of input records.

The network observes the connection among enter information and the preferred final effects and learns to imitate the rule. While studying with a educate you typically want to address a couple of values: a pattern input signal and a desired output (required reaction) of the community to the enter sign. Of route, a network may also have many inputs and loads of outputs.

The pair, in fact, represents a complete set of input facts and output records that need to paintings as an entire solution for a project.

The components (the statistics for a task and the output answer) are typically required. The instructor and train phrases require factors at this factor. A display is not usually a man or women who teaches a community, regardless of the truth that people work with and educate networks.

In exercising, the placement of a train is taken over by means of a computer that models the precise network. Unfortunately, neural networks are not very smart. Effective getting to know a hard challenge requires lots, or from time to time even masses of masses, of steps! No human might have the electricity and staying energy to educate a device that learns so slowly.

That is why a trainer, or train in this context, refers to a computer software, provided by means of manner of a human, with a so-referred to as getting to know set. What is a analyzing set? Here is a desk displaying sample facts concerning pollution prices in diverse US towns. Any extremely good type of statistics is probably used to provide an cause at the back of this concept. However, it's far crucial to use real-life statistics.

In a actual database, I defined that, after leaving the elements of the precise window (a software going for walks in this database)-- the statistics accrued within the database, as you could see--we're capable of isolate those with a view to be used as outputs for the community.

Check out the sort of columns of the table with the arrow at the bottom of the determine. The information will allow us to are searching forward to many levels of air pollutants. The statistics covers population figures, industrialization ranges, weather situations, and different factors. When the ones facts are used as inputs, the community will should assume the not unusual diploma of air pollutants in each metropolis. For a town in which pollutants stage facts has no longer been compiled, we're capable of need to wager.

That is wherein a formerly educated community will visit paintings. The gaining knowledge of set facts—known pollution facts for numerous cities—has been placed in the appropriate column on the desk, this is marked with a pink arrow (output).

Therefore, you have got exactly the fabric you need to educate the network: a fixed of facts pairs containing the appropriate input and output statistics. We can see the motives (population, industrialization, and weather situations) and the result (air pollutants charge).

With this technique, the community uses the ones statistics and will learn how to function nicely (estimating the values of air pollutants in towns for which proper measurements have not but been made).

Exemplary mastering techniques may be thoroughly cited later. In the interim, every other element of that. The letters in a unmarried column of the desk are slightly seen because of the truth they seem in gray in location of black.

This shading indicates that the data portrayed are sincerely much less essential. The column includes the names of specific cities.

Based on this facts, the database generates new records and outcomes, however, for a neural community the statistics inside the columns is useless. The diploma of air pollutants isn't always related to the call of a city, so, no matter the fact that the ones names are available within the database, we might not use them to teach networks.

Databases regularly incorporate hundreds of statistics records that isn't needed to train a community. We must do not forget that the train involved in network reading will typically be a fixed of facts that is not used "as is," but is adjusted to function as a getting to know set following careful selection and right setup (data for use as inputs and facts to be generated as outputs).

A community want to not be plagued thru statistics that someone is aware about isn't useful for checking answers to a particular trouble.

Besides the technique of the getting to know with a teacher described earlier, a series of techniques of learning without a teacher (self-studying) is likewise possible. The strategies simplest include passing a series of check statistics to the enter of networks, with out counting suitable or even expected output alerts.

It appears that a properly designed neural community can use simplest observations of the the front signals to assemble a sensible set of guidelines of its very private interest primarily based on them, most customarily counting on the truth that classes of repeated (likely with best variety) enter indicators are routinely detected and the network learns (spontaneously, without any open learning) to apprehend those not unusual kinds of indicators.

A self-gaining knowledge of network requires a reading set which incorporates information provided for enter. No output data are supplied because, on this approach, we need to make easy the expectancies from the network regarding the analysis of precise facts. For example, if we have a look at the information in Figure three.1 to analyzing with out a teacher, we might use simplest the columns described as input facts, in preference to giving the community statistics from the column indicated via the use of the crimson pointer .

A self-studying network couldn't be capable of count on the stages of pollution in precise towns, as it cannot advantage know-how on its personal. But via studying the records on specific cities, the community can also pick out (with none assist) a set of massive business organization cities and research to distinguish them from small u.S. Towns that lie inside the center of agricultural regions.

The community will develop this distinction from the given enter information by means of using following the rule of thumb indicating that commercial enterprise towns are comparable, and agricultural towns moreover percent many common houses with every distinct.

44

Neural networks may (with none assist) use a rule to differentiate those cities with right and lousy weather and decide extremely good classifications, relying fine on values of decided enter facts. Notice that the self-gaining knowledge of network is probably very thrilling from at the same time as comparing such community sports activities the activities of human brains.

People also have the capability to spontaneously classify the devices and phenomena they encounter. After a suitable class has been achieved, people and networks understand some different object as having traits belonging to a previously identified magnificence. Self-studying is very thrilling, counting on the usage.

It calls for records that would be inaccessible or hard to gather. A neural community will collect all wanted statistics and information segments with out out of doors help. Now, you may consider (for amusing and for stimulating your creativeness, in vicinity of from actual need) that a self-analyzing network with a tv digicam is despatched in an unmanned area probe to Mars.

We don't know the situations on Mars . We do now not know which devices our probe should understand or how many education of items can be positioned! But even with out that facts, the probe will land, and the community will begin the approach of self-gaining knowledge of.

At first, it acknowledges nothing and simplest observes its surroundings. However, over time, the process of spontaneous self-organisation will allow the community to learn how to discover and differentiate numerous sorts of input signs: rocks from stones and plant paperwork from incredible living organisms.

If we deliver the network sufficient time, it's going to learn how to differentiate Martian guys from Martian ladies, irrespective of the truth that its author did now not understand that Martian people existed! Of route, this self-gaining knowledge of probe on Mars vehicle is a hypothetical introduction despite the truth that networks that create and understand numerous patterns exist and are in common use. We is probably interested by figuring out how many sorts of a little identified sickness can be decided in reality. Is a state of affairs one infection unit or severa? How do the components vary? How can they be cured?

It is probably enough to construct a self-getting to know neural network to preserve the facts on registered sufferers and their houses over an extended duration.

This network will yield information on what number of not unusual groups of signs and symptoms and symptoms and signs and symptoms were detected and which necessities may be used to categorise patients into wonderful corporations. Applications of neural networks to goals like the ones would likely even cause a Nobel Prize! This method of self-learning, absolutely, has many issues, which we are able to describe later.

But self-analyzing undeniably has many benefits. You is probably amazed to investigate that this device isn't as famous because it need to be, given its many packages.

Methods of Gathering Information

Let's look at the manner of reading with a trainer. How does a network gain and collect information? Keep in thoughts, every neuron has many inputs, via which it receives the symptoms from unique neurons and from network facts to feature to its processing outcomes. The parameters, referred to as weights, are combined with access information. Each input signal is first modified by the burden ,and most effective later introduced to the alternative signals. If we change the values of the weights, a neuron will start to feature inside the community in a cutting-edge manner, and in the end the entire community will paintings in a one-of-a-type way. A community's analyzing capacities is primarily based upon on the choice of weights, so that all neurons will carry out the appropriate duties demanded by means of the network A network may also additionally comprise thousands of neurons, and every one in all them may additionally additionally cope with loads of inputs, so it's is impossible for these varieties of inputs to create the important weights simultaneously and without direction.

We can, but, format and collect analyzing through using beginning community sports activities with a splendid random set of weights and step by step enhancing them. In every step of the mastering technique, the values of weights from one or numerous neurons undergo modifications. The guidelines for trade are set in this kind of manner that all of us neuron can qualify which of its personal weights ought to change, how (thru being prolonged or reduced), and what kind of.

The instructor passes on the data approximately the important adjustments inside the weights that may be utilized by the neuron.

Obviously, what could not change is the truth that the method of changing the weights runs thru each neuron of the community, spontaneously and independently. In reality, it may rise up with out direct intervention through the person supervising this manner.

What's extra, the machine of mastering of one neuron is impartial from how some special neuron learns. Thus, gaining knowledge of can arise simultaneously in all neurons of a network (of direction, this could best stand up in a suitable network with an ok virtual tool, and no longer thru a simulation software). This characteristic allows us to reap very excessive speeds of learning and a mainly dynamic boom of skills of a community, which actually grows an increasing number of clever earlier than our eyes! We have to stress a key factor all yet again : a trainer need no longer get into the facts of the technique of learning. It's sufficient for the instructor to provide a network an example of a accurate solution. The network will compare its private solution, obtained from the example from the gaining knowledge of set, with the solution that changed into recorded as a model (maximum probably correct) in the studying set .

Algorithms of studying are built just so the records approximately the price of an errors is enough to allow a community to accurate the values of its weights. Every neuron corrects its personal weights on all entries, one after the other under the manipulate of the right set of rules, after it receives an mistakes message.

It depicts a smooth however green mechanism. Its systematic use reasons the community to perfect its very very own sports sports, till it is finally able to clear up all assignments from the learning set, at the grounds of generalization of this know-how. It also can manage assignments that it's far going to be introduced to on the examination degree. The manner of network studying described in advance is used most customarily, however some assignments (e.G., image reputation) do no longer require a community to have the exact value of a favored output signal.

For green gaining knowledge of, it's miles enough to present community tremendous current statistics on a topic, whether or not or now not its cutting-edge conduct is correct, or no longer. At instances, network experts communicate about "rewards" and "punishments" with reference to the way all neurons in a network discover and introduce proper

corrections to their very very very own sports without outdoor course. This analogy to the schooling of animals isn't always unintentional.

Organizing Your Network

The style of values of the weight coefficients in each neuron is counted based mostly on particular hints (paradigms of networks). The numbers and styles of the policies which can be used nowadays are immoderate due to the fact maximum researchers try to characteristic their very very very own contributions to the vicinity of neural networks as new regulations of studying.

We can now maintain in thoughts two fundamental policies of gaining knowledge of with out using mathematics: the rule of thumb of the fastest fall, that is the concept of most algorithmic studying with a trainer, and the Hebb rule, this is simplest example of studying with out a trainer.

The rule of the fastest fall is predicated on the receipt, by way of way of each of the neurons, of signs from the network or from distinct neurons. The signs give you the cease result from the primary degrees of processing the records. Neurons generate the output sign the usage of their information of the earlier settled values of all amplification factors (weights) of all entries and (maybe) the edge. In the closing section, we stated many strategies of marking the values of output indicators by using the use of neurons primarily based on enter sign.

At every step of the method of gaining knowledge of, the rate of the output sign of a neuron is in comparison to the instructor's answer within the gaining knowledge of device.

In that duration of divergence, which takes area generally at step one of the mastering method, the neuron attempts to discover the difference among its very private output sign and the fee of the signal that the instructor indicates is correct. The neuron can then decide a way to exchange the values of the weights to lessen the mistake.

It may be useful to apprehend the place of an mistakes. You already recognize that the pastime of a network is based totally mostly on the values of the load coefficients of the constituent neurons. If you apprehend the set of all weight coefficients taking place in all neurons of the neural network, we're able to apprehend how a community can act.

In unique, we are capable of hypothesize a community of assignments, examples and answers which can be handy as part of a studying set. Each time the network gives its non-public solution to a question, you could evaluate its method to the right solution found within the studying set, due to this revealing the network's mistakes.

Measuring the mistakes is typically the difference between r the neural network's answer and the fee of the bring about m the getting to know set. To estimate the overall sports activities sports of networks with defining gadgets of weight coefficients within the neurons, you commonly use the overall of the squares of mistakes collected through the usage of the network for every case from the studying set. The mistakes are squared in advance than addition to keep away from the trouble of mutual reimbursement of fantastic and terrible errors. This consequences in heavy consequences for huge errors.

Thus, a times massive mistakes will yield a quadruple thing inside the trendy end stop end result. Each u . S . Of extremely good and awful mastering of this neural community may be joined at the factor at the horizontal (the slight blue) floor tested within the determine, with its weight coefficient coordinates.

Just consider it: now you have localized those weight values inside the neural network that study the location of the red issue inside the decide.

Examining a neural network through way of the use of which means all factors of the studying set will treatment the whole rate of the errors. At the red element, you can vicinity a pink arrow pointing upwards.

The top will constitute the final price of the mistake based totally at the vertical axis. To do the equal steps the use of the blue pointer, just consider doing the identical acts for all combinations of coefficients — that is, for the all points of the blue moderate.

We will see that many mistakes can be large, and others, smaller. If you had the staying power to test your community commonly, you can see the errors' ground placed over the only-of-a-kind weights.

You can see lots of them at the floor. This is the community giving severa mistakes, and may be avoided. A neural community giving small errors is likewise viable.

Usage of Momentum

One technique of maximizing the getting to know pace without affecting stability is thru the use of an additional element referred to as momentum inside the set of guidelines of reading. Momentum makes the technique of reading broader via changing the weights on which the approach is based totally upon and the modern-day errors, and lets in reading at the first step.

It allows for a assessment of analyzing with and with out momentum and the way of changing the burden coefficients. We can show pleasant two of them, and accordingly the drawing ought to be interpreted as a projection at the aircraft decided by using the use of manner of the weight coefficient, w_i, and the burden model gadget, w_j, that takes vicinity within the n dimensional area of the weights.

We can see the behavior of high-quality inputs for a selected neuron of the network, but the techniques in unique neurons are similar.

The pink (darkish tone) factors represent beginning elements (the placing earlier than the begin of reading the values of weight coefficients). The yellow elements mean the values of weight coefficients received inside the course of the getting to know steps. An assumption has been made that the minimum of the error function is attained at the factor indicated by using manner of the plus sign (+).

The blue ellipse indicates the outline of the strong mistakes (the set of values of weight coefficients for which the technique of studying maintains the same degree of mistakes). As established within the determine, introducing momentum stabilizes the mastering approach because the weight coefficients do now not change as violently or as often.

This will make the method more nicely sufficient because the consecutive points the way to the first rate factor faster. We use momentum for learning from a rule as it will enhance the ratio of correct solutions received, and the execution prices are decrease.

Chapter 4: Using Neural Networks

Using Neural Networks in a Practical Way

This will lead you to a domain that because it should be describes, step by step, all of the required obligations for the usage of those programs legally and for free of charge. You'll need those programs to verify the information about neural networks on this e-book. Installing the ones programs for your computer will will let you find out severa competencies of neural networks, conduct interesting experiments, and in reality have amusing with them.

Don't fear. Minimal enjoy of putting in applications is wanted. Detailed installation instructions appear at the internet internet site. Note that updates of the software program software utility will fast make the data in this e-book antique. Despite that opportunity, we should provide an motive behind some of the strategies.

Downloading the packages from the internet web site on-line can be very easy, and can be finished with one click on on of the mouse.

However, acquiring the applications is not sufficient. They're written in C# language and want to be hooked up libraries in .NET Framework (V. 2.Zero). All the required set up software program is at the internet site on-line.

The first step is installing the libraries. You may additionally have some of the ones libraries set up to your computer already, so that you may also find out this step useless. If you're about this, you can pass the .NET Framework step, but we advocate you do it actually in case.

If the right packages have already been hooked up on your pc, the installation detail will decide that it would no longer need to install something . However, the internet web site can also additionally moreover contain more moderen variations of the libraries, wherein case the installer will go to art work. It's constantly smart to replace antique software software program with more recent variations. The new software software will serve many functions beyond the ones described in this ebook.

After you placed within the important .NET Framework libraries, on which further movements rely consistent with web web page guidelines, the installer will run again. This will will permit you to set up all sample applications automatically and painlessly.

You'll want them for appearing the experiments included on this e-book. When the set up is completed, you will be able to get proper of get right of entry to to the packages via the start menu used for most programs. Performing .NET experiments will assist you recognize neural networks theoretically, in addition to exhibit their use for sensible capabilities.

If you do no longer take transport of as true with this save you is easy, that is the hard a part of set up: the installer asks the character

(1) wherein to position within the applications and (2) who've to have get proper of get admission to to.

The remarkable reaction is to preserve the default values and click on on "Next". More inquisitive clients have more alternatives: (1) downloading the source code and (2) installing the integrated improvement surroundings called Visual Studio.NET.

The first alternative allows you to down load textual content versions of all of the pattern programs whose executable variations you'll use even as analyzing with this e-book. It's very interesting to appearance how those packages were designed and why they artwork as they do. Having the deliver code will allow you—if you dare—to adjust the applications and carry out even greater interesting experiments. We need to emphasize that the supply codes are not crucial in case you absolutely need to apply the programs. Obviously, having the codes will let you look at greater and use the to be had property in interesting strategies. Obtaining the codes isn't some of paintings.

The second opportunity, installing Visual Studio.NET, is for ambitious readers who want to alter our programs, layout their very personal experiments, or write their personal packages. We inspire readers to put in Visual Studio.NET although they simplest need to view the source code. Making viewing the code an awful lot less hard will most effective take a few more mins. Visual Studio.NET may be very easy to install and use.

After installation, you will be capable of carry out many more responsibilities, together with such as extern devices, diagnosing programs with out trouble, and rapid producing complex versions of the software program.

Remember that installing the supply codes and Visual Studio.NET are completely non-obligatory.

To run the sample applications to be able to assist you to create and check the neural community experiments defined on this ebook, you need simplest set up the .NET Framework (step one) and sample applications (the second step).

What's subsequent? To run any instance, you need to first pick out out the proper command from the Start/Programs/Neural Networks Examples menu. After making your choice, you could recreate and observe every neural community defined on this book along side your laptop. Initially your machine will use a network whose shape and measurements were designed through means people. After you benefit get right of entry to to to the deliver code, you may be capable of regulate and exchange a few component you want. The preliminary software program will create networks stay in your pc and will let you teach, check, have a look at, summarize, and take a look at them. This technique of coming across the abilties of the neural networks—thru constructing and making them artwork—could in all likelihood additionally be far greater pleasing than getting to know the idea and attending lectures.

The way a network works is primarily based upon on its structure and purpose. That is why we're able to speak approximately precise conditions in next chapters. We will begin with picture popularity, because it's the handiest network characteristic to provide an explanation for. This type of community receives an image as an input and, after categorization of the photo based totally on previous studying, it produces an output.

This shape of community have become referred to within the last bankruptcy. A network handles the task of classifying geometric pictures through figuring out discovered and handwritten letters, planes, silhouettes, or human faces.

How does this form of community function? To solution that, we must start from an absolutely simplified network that consists of simplest one neuron. "What? You preserve in thoughts that a unmarried neuron can not shape a network, due to the reality a community want to include many neurons all associated with each other? The quantity of neurons might not really keep in mind, and so even a small network can produce exciting outputs.

The Capacity of a Single Neuron

As stated in advance, a neuron gets enter signals and multiplies them via factors (weights assigned in my view all through the studying method), which may be then delivered collectively and blended proper proper into a unmarried output sign. To recap, you apprehend that summed signs in greater complex networks combine e to yield an output sign with the right feature (typically nonlinear). The conduct of our simplified linear neuron consists of a protracted way much less hobby.

The charge of the output sign relies upon on the extent of splendor between the signs of every enter and the values in their weights.

This rule applies preferably high-quality to normalized input alerts and weights. Without particular normalization, the rate of an output signal may be treated because the size of similarity among assembly of the enter signals and the meeting in their corresponding weights.

You can also say that a neuron has its very own memory and stores representations of its statistics (forms of input records as values of the weights) there. If the input signs wholesome the recalled pattern, a neuron recognizes them as acquainted and solutions them with a robust output signal.

If the community reveals no connection the various enter indicators and the sample, the output signal is close to zero (no popularity). It's feasible for a complete contradiction to arise a number of the enter signals and the weight values. A linear neuron generates a horrible output signal if so. The extra the contradiction a few of the neuron's picture of the output sign and its actual rate, the more potent its horrible output.

We encourage you to install and run a clean software named Example 01a to perform a few experiments. You'll study even greater approximately networks if you attempt to beautify this system.

After initializing Example 01a, you may see a window within the software. The textual content in the top section explains what we are going to do.

The blinking cursor signals that this device regarding flower characteristics is searching out the burden of the neuron's enter (in this example, the aromatic charge). You can enter the cost via typing a number of, clicking the arrows next to the sector, or using the up and down arrows at the keyboard. After putting the fee for the fragrant feature, circulate without delay to the following area that corresponds to the second function L coloration.

Let's count on that you want your neuron to choose colourful and fragrant flora, with more weight for coloration. After receiving the tremendous answer, the window of this machine will look like this application, and every particular one you could use allows you to exchange your desire and pick out another enter. The utility will try to update the effects of its calculations.

After we input the feature facts (which ,in reality, are weight values), we can observe how the neuron works. You can enter numerous units of facts, as proven in Figure 4.4, and this system will compute the subsequent output signal and its which means. Remember that you may trade the neuron's options and the flower description at any time.

If you're the use of a mouse or arrow keys to go into facts, you do now not ought to click on the "recalculate" button on every occasion you want to peer the stop quit result; calculations are finished routinely. When you enter a whole lot of from the keyboard, you want to click the button due to the truth the computer does no longer apprehend

in case you finished entering into the variety or left to move get a tuna sandwich.

The subsequent stage is to test with a neuron in an uncommon situation. The factor of the test in this window is to observe how the neuron reacts to an item that differs from its first-rate colorful and aromatic flower. We confirmed the neuron a flower complete of colours without a fragrance. As you may see, the neuron liked this flower too!

You'll find out "gambling" with the Example 01a software program application to be a profitable workout. As you enter diverse facts devices, you may speedy see how a neuron features consistent with a easy rule. The neuron treats its weight as a model for the input sign it desires to recognize. When it gets a aggregate of indicators that corresponds to the load, the neuron "well-known" some element familiar and reacts enthusiastically with the useful resource of manner of manufacturing a robust output sign. A neuron can sign indifference thru a low output sign, or perhaps advise aversion via a horrible output, due to the truth its nature is to react absolutely to a signal it recognizes.

Careful exam will display that the behavior of a neuron is based upon great on the perspective the diverse vector of the weight and the vector of the input signal. We will use Example 01b to further reveal a neuron's likes and dislikes with the resource of way of imparting an terrific flower as a point (or vector) inside the enter location.

When you place the alternatives of a neuron, you tell it, as an example, to decide on fragrant and colorful vegetation. Fragrance and color are separate weight vectors. You can draw axes. On the horizontal axis, you can be aware the values of the primary feature (fragrance), and suggest the values of the second

function (coloration) on the vertical axis. You can mark the neuron's choices at the axes. The point wherein those coordinates meet indicates the neuron's alternatives.

A neuron that values nice the perfume of a flower and is detached to the shade might be represented through a factor placed maximally to the proper (a excessive cost for

72

the number one coordinate), but at the horizontal axis, may be set at a low quantity or 0. A puttyroot flower has stunning colorations and a susceptible an on occasion unsightly smell. The puttyroot can be located excessive at the vertical axis (immoderate colour rate) and t left at the horizontal axis to suggest a inclined or unsightly fragrance. Flower color is valued at the vertical axis and fragrance, on the horizontal axis. You can cope with any item that you'd like a neuron to mark with this technique.

Questions

1. Using the .Net framework, put into effect a neural network.

Chapter 5: Profiting From Neural Networks

Neural networks are top notch if you need as a way to create diverse matters which may be automated inner a computer or a way. They are even higher, in spite of the truth that, at the same time as you are able to take the time which you want and begin to take advantage of them. Profiting out of your neural networks is a lot greater than truely being able to add the one of a type options on your laptop structures. You can now use neural networks if you want to make huge income within the stock markets or even forex.

Now that neural networks are becoming extra well-known and they'll be able to learn to be self-taught, you may use them for greater topics than what you were ever able to do inside the beyond. Your neural networks want to be adjusted as a way to attain remarkable levels and also you ought to make certain that you are going so one can encompass all the one among a kind cash-making talents in with the facts you take a look at them.

When you realize a manner to make a neural community, you have got greater than only a potential – you have got a profitable information that you may use to assemble capital and create financial freedom for your self.

Computers over Brains

Human brains are high-quality due to the fact they may be intuitive but neural networks are able to grow to be nearly as intuitive due to the fact the thoughts. When they will be taught to be self-taught inside the right way, they may be able to make the high-quality selections near the alternatives which can be included with neural networks.

In the beyond, humans have been the popular method that have been traditionally used for purchasing and promoting. While a few computers were capable of help trade on an intuitive degree, they had been now not able to select up at the equal matters that humans can.

With neural networks, people do now not really need to be worried within the shopping for and promoting approach.

Specific Money-Making Definition

The capability of a neural community to make coins is absolutely counting on its capability with a reason to research the numerous topics that the administrator has taught the network that permits you to do. It is essential to make sure that you have end up the excellent revel in feasible along with your neural community with the useful resource of using on the aspect of the whole lot which you want it to do. If you set up your community to exchange and discover ways to engage with the numerous elements of the stimuli which is probably going into it with the buying and promoting quarter, you may be able to make coins from it.

Training Through Data

The notable way to get your community to make cash when you are doing numerous matters with it's far to push various types of records thru it. This will permit the pc to essentially "observe" the facts and learn how to have interaction with it. It is a schooling approach that the neural community wishes that permits you to go through in precise regions that it is in. You need to make sure that you are doing what you may so you can get the neural network adjusted to your buying and selling competencies.

Rules Created for Networks

When you are strolling with the neural network and you are pushing the facts thru, you may want to installation a few pointers at the begin of the shopping for and promoting way. This will imply that you'll be looking to pay near hobby to the diverse topics that it's miles doing and the manner that remarkable sorts of stimuli are pushing thru it. While you are looking it, you want to trade the hints that you set for your neural network.

Working them for Monetary Gain

As you're studying more about neural networks and the numerous subjects that pass on with it, you may studies the manner that you can make coins from them. Money may be had in plenty of brilliant techniques via neural networks. The maximum famous techniques are through growing them for unique entities and through trading with them. You can also invest in neural networks however the move again on that takes quite a long term and is frequently not definitely well well worth the time that you need to wait to see on it.

Trading System

There is an entire buying and selling tool that is associated with neural networks. People can buy, alternate, promote and circulate across the unique networks that they've. The wonderful entities which may be a part of the neural network buying and selling area will assist to determine the amount this is covered with the shopping for and promoting. When you're running in neural networks, it's far worth looking into the exchange region that comes along aspect it – a few people can pay a whole lot of coins for the advent of these networks whilst others can pay to exchange the specific additives of the network that they have created. While the buying and selling performs a massive feature within the buildup of capital on the aspect of your neural network, it's going to moreover help to create a experience of network in the community.

Investing Through Networks

The funding which you make right into a neural network on the equal time as building it is also made from it gradual. There are usually no longer very high costs which are related to neural networks however you want to commonly be organized for the fees which you might likely incur at the same time as you're using the neural community. You can, as an opportunity, invest in the real neural community company. You can do that in case you need to get returns to your searching for and promoting and in case you are going as a way to increase the amount of coins that you have for neural making an funding.

Neural Network Architecture Search

"Neural Architecture Search: A Survey" specializes in constructing this form of neural network that might evolve on its very very own, create offspring neural networks and discover the maximum superior neural community layout for any given hassle. The neural shape are seeking (NAS) is in essence computerized device learning that includes three parameters: are seeking for region, search approach, and standard performance estimation. The most effective human involvement is placing these 3 parameters and assessing the consequences, despite the fact that the paper laments the fact that human beings introduce their bias via restricting the neural network's seek area.

Search place refers back to the kind of form fashions we'll allow the neural community inspect and recombine to attain some issue virtually new and higher than any of them. A novel proposition in terms of are seeking area is the creation of "cells", hand-crafted neural community additives which can hold the information multidimensionality or be made to lessen statistics to a vector. Cells can be transplanted to some different neural network or stacked on top of each distinctive for top notch overall performance income, but the belief is to allow the neural community integrate them in arbitrary strategies until there can be big progress.

Any kind of are searching out strategies may be employed thru the neural network, regardless of the fact that the general overall performance of some of them declines because the community scales, making them unfeasible. For example, reinforcement studying are seeking approach utilized by Zoph and Le in 2017 required 800 computer photographs playing cards on foot for three-four weeks. However, evolutionary techniques have been used all the manner yet again inside the early 90s and allowed amazing seek location exploration with low-key resource intake. The primary concept inside the back of the evolutionary technique is that the figure neural network runs for some time until it reaches a certain form, notes its standard overall performance on the venture after which passes the layout on to its offspring, a brand-new neural network that now has some concept on how nicely sure cells perform on the project and can weight them therefore. By frequently removing notable cells and their layouts the offspring can in the end obtain the last neural community structure for the given project. Now, all that's left is overall performance estimation.

Despite lowering corners with cells and the use of the evolutionary approach to increase scalability NAS still incurs big useful resource drain even as it's time to test the proposed neural network structure, prompting the authors to hotel to performance estimation that is, in essence, lo-fi measurement. The trick is that the estimate can't oversimplify or the effects may be no nicely, so the authors motel to without a doubt rating resulting neural networks based totally mostly on their normal overall performance, with the concept that we don't need to understand how well the awesome one is, simplest that it is the splendid one of the bunch.

The authors finish that NAS fared well, however ranking performance is even though tough because of the dearth of not unusual benchmark requirements in deep mastering. Another observation is that NAS doesn't virtually monitor why a high-quality shape performs in a high nice manner and that knowledge cell groupings (moreover called "motifs" in the paper) may additionally provide perception into how neural networks art work.

Noise recognition

The 2018 research paper "Noise Adaptive Speech Enhancement Using Domain Adversarial Training" seems at speech reputation neural networks expert the use of deep mastering and their ability to cope with audio resources with additional history noises not encountered at a few level within the training phase. Rather than compiling the list of each noise ever created, the paper indicates using domain adverse training (DAT) to educate extra subroutines of the neural network: a discriminator that attempts to decide if the noise is coming from the unique audio supply and a feature extractor that attempts to offer the super noise to confuse the discriminator. By pitting a discriminator against a function extractor over an arbitrarily long time period, the DAT method permits the evolution of a specialised discriminator module that may later be merged with the audio popularity neural community for 26-55% better overall performance in gadgets collectively with cochlear implants and speech-to-text software application software. This shape of hostile education is wellknown for neural networks because it allows scientists to leverage the blazing pace of pc systems for green reading in evaluation to what could appear within the event that they were manually fed examples.

Scene recognition

"From Volcano to Toyshop: Adaptive Discriminative Region Discovery for Scene Recognition" seems at an photo classifying neural network this is first professional to label gadgets within the scene, then label the scenery itself, and ultimately contextualize every to attain at an outline of the area, for example, "paintings college" or "campsite". The concept is to teach the neural network to understand high-quality gadgets as as a substitute deterministic of the region, as an instance, locating a tent within the image strongly suggests it's outside. Such neural networks exist already however are computationally demanding each in education and in operation; deep mastering allows the operator to set an arbitrary amount of signifiers within the photo to gain scalability.

Decrypting hidden messages

Steganography is deliberately hiding facts in exceptional records, and steganalysis is a way to unveil such hidden messages, every of which neural networks do lots higher than humans as observed in "Invisible Steganography via Generative Adversarial Network". The idea is to teach separate neural networks, one in hiding the records and the alternative in unveiling it, through pitting them toward one another. The test given turn out to be to have one neural community have a look at the cover image, find the most appropriate pixels, successfully cover a gray image proper right into a shade one, and then send it to the opportunity neural network for evaluation. When strolling in tandem the two networks allow, for instance, the headquarters to transmit a mystery message encoded in a undeniable photo via public channels to operatives inside the issue who can decode it the use of the opportunity part of the duo.

Automatic query technology

High school quizzes taught us many information, collectively with that mitochondria is the powerhouse of the cell. Preparing the quizzes required loads of touchy but hasty paintings that might pass wrong in any variety of places, making university university college students experience cheated out of a awesome grade due to a malformed question. With the help of deep studying, we might be getting ready to an age wherein questions and solutions are unmistakably made from swaths of textual content through a neural community, the powerhouse of the mastering manner.

"Improving Neural Question Generation the use of Answer Separation" seems at routinely growing questions and answers out of any amount of text, starting from unmarried sentences to massive paragraphs. The purpose isn't to have the neural network create a difficult draft of the quiz however an actual viable version from scratch that doesn't want any proofreading or enhancing.

This is completed through manner of having the neural community become privy to and masks the answer with a token signifier and semantically conclude the proper phrase series and pronouns.

For instance, the sentence "John Francis O'Hara emerge as elected president of Notre Dame in 1934" consists of 3 questions: who (John), what (president) and at the same time as (1934). By masking each of the 3 solutions, we educate the neural network the manner to pose a query and check it in opposition to the masked answer; by way of way of adding an interest mechanism we offer higher weight to key phrases and statistics tidbits, mimicking the way people parse questions and hold information. This technique permits extraction of maximum value out of the identical textual content but also acts as an anti-cheat measure at some point of the quiz itself for the purpose that college college college students can't duplicate an answer from someone else.

The neural community turn out to be examined the usage of 23,215 text samples originating from 536 text sources with about 100,000 questions and solutions created manually for actual quizzes. The results showed that fine 0.6% of questions created with the resource of this neural network erroneously discovered out the complete solution and 9.5% gave a hint as to what the solution have come to be, each being commonplace prone factors of neural networks handling growing quiz questions. "What", "how" and "who" were the 3 maximum not unusual pronouns that the neural network guessed efficiently, despite the fact that the not unusual precision for distinctive question kinds wasn't so stellar; the authors attributed this to fifty 5.Four% of all questions having "what" and one-of-a-type pronouns not being nearly as represented within the schooling dataset.

three-D seen reconstruction of 2D gadgets

Great painters apprehend the manner to apply sunglasses and attitude to make the canvas a actual window into some other global. Even despite the truth that on a few degree we apprehend that this form of painting is an illusion, our thoughts snaps the three-d photograph collectively and offers it as actual, filling within the blanks using what it's miles aware of approximately the outdoor global. It seems neural networks are in a position to a few aspect similar and may reconstruct the unseen aspect of an object primarily based sincerely off of definitely one in every of its 2D views.

"Deep Learned Full-three-D Object Completion from Single View" is a joint USA-Italy research paper looking at how pixels may be end up voxels, superb described as volumetric pixels. The most vital reason of this check is to decrease the type of views wanted to complete a 3-d object, in all likelihood to be used with robots on a decent computational budget transferring through an environment and interacting with real items.

Since the authors wanted at the least one view, they determined to go along with that and ended up in reality succeeding.

The neural network modified into knowledgeable the usage of 5,000 model presets from CAD, a well-known modeling software, each model having eight snapshots taken from precise angles to reach at 40,000 demanding situations. All models had been at a 30x30x30 resolution, which supplied a task even as it came to preserving all the nuanced abilties of a version, however the neural community controlled to attain brilliant consequences although, restoring ninety % of the precise model.

Rain streak elimination from an photo or video feed

Setting up a remotely handy outside virtual digicam can sound like a fun check; that is till the rain falls and rain streaks make it appear like we're peeking via a white curtain. The real purpose of this impact is that raindrops have a immoderate tempo on the same time as reflecting moderate, causing white streaks to seem on any picture taking photographs device. This devalues all top notch system getting to know strategies that depend upon picture processing, alongside aspect facial popularity, and as a end result getting rid of rain streaks becomes a top priority venture. There currently exist numerous de-raining applications, a number of which use system learning, but they usually smash the photograph first-class both with the aid of blurring the ancient past or thru way of ruining image evaluation. "Rain Streak Removal for Single Image through Kernel Guided CNN" shows the usage of a neural network referred to as KGCNN to clean the pictures up with the desires being to preserve as a bargain picture awesome as possible and do away with rain streaks from the picture in a computationally light-weight way.

KGCNN could make the most a acknowledged assets of raindrops, that being that they spark off a small however perceptible quantity of movement blur. Knowing the general path of a raindrop, in particular that they typically have a tendency to collapse, can assist us assemble KGCNN to be able to deconstruct any picture feed proper right into a historic beyond and texture layer, with the goal being to discover movement blur inside the latter and use this information lower again on the composite photo to block out the rain streaks. The history layer consists of everything except the rain, and the feel layer consists of quality the raindrops, making it smooth to pick out at a glance if KGCNN works as anticipated.

Computer systems intrusion detection

In sci-fi books such as William Gibson's Neuromancer, hackers plug into the net via a literal plug inserted into the lowest of the cranium and jockey the software program round; intrusion detection is completed by means of using an AI known as Black Ice that guards proprietary our on-line world and attempts to fry the hackers' brains. Hackers and intrusion detection exist proper now in a much greater prosaic way, however neural networks promise that as a minimum that latter element is about to end up a bargain greater interesting.

"Statistical Analysis Driven Optimized Deep Learning System for Intrusion Detection" investigates the growing risk of realistic malware and hacking assaults that could jeopardize banking structures, energy grids or health facility document databases. It's not sincerely that the whole thing is networked, however the sheer length of such networks makes updating a nightmare and will boom their attack ground, leaving them uncovered to any hacker with an opportunity; it's as easy as walking to the sort of terminals linked to the intranet, the inner network, and popping in an inflamed USB. It's probably how WannaCry ransomware hit sixteen UK hospitals in May 2017, locking all clinical documents in the again of a paywall and perilous deletion except $three hundred in ransom modified into paid in Bitcoin.

This studies paper shows a scalable, slight-weight way to hold big networks steady, no matter whether their additives are updated or not, through the usage of using neural networks that sift via a massive amount of data to assume intruder behavior and deny them get entry to. We understand from different clinical areas that neural networks can acquire near-human conventional usual overall performance in times of person reputation and 3-D object reconstruction, so it's of superb interest to find out a sustainable, cheap possibility to antivirus packages and old control get proper of access to sports that motive limitless grief to guide staff, including usernames and passwords.

The neural network assigned to network safety does information preprocessing to put off outliers, characteristic extraction to discover commonalities between customers, and class to differentiate amongst benign and malign clients. In fashionable, intrusions come as: probing that scouts the purpose community for weaknesses and open ports; denial-of-carrier, which serves to incapacitate goal community and estimate its skills; consumer-to-root that is meant to benefit root get right of access to; and root-to-neighborhood that is supposed to perform operations on a community gadget after root get right of get right of entry to to has been gained.

Neural community was first professional the use of one hundred twenty five,973 samples belonging to any of the 4 intrusion commands and examined with every other 22,544 samples, resulting in an accuracy of 77.Thirteen% for probing attacks, ninety seven.08% for denial-of-provider, 87.10% for person-to-root however notable 11.Seventy 4% for root-to-close by. These numbers propose that every one protection functions need to recognition on preventing get right of get right of entry to to to root systems, which may be the ones which could hassle commands to subordinate machines, which include the whole community or individual terminals; as quick as the inspiration has been hijacked, every compromised device, or perhaps the whole community, is decisively inside the arms of the attacker, as seen with WannaCry ransomware.

Logical thinking

Logic units man aside from amoebas and we may want to us see each ourselves and amoebas with a high degree of reality. The capacity of logical reasoning derives from a symbolic illustration of the arena and is the most yearned-for college scientists need their neural networks to have. It's now not just any sort of nicely judgment, despite the reality that, but a unique type referred to as ontological common revel in that deals with how we came to be and what ties us together. Ontological wonderful judgment can be applied to global locations, people, animals, trees, rocks or some different viable cloth or metaphysical entity to seize the cloth of time and remedy it lower lower back to its place to begin. As neural networks get set to compete in competition to human beings in all fields of existence and technological expertise, they'll slowly but in reality discover ways to navel-gaze just like the relaxation parents with the loose time to achieve this.

"Ontology Reasoning with Deep Neural Networks" appears at how a neural community that's been given data approximately human beings draws conclusions approximately their relationships; as an instance, two separate human beings that display as much as be parents to the identical individual should be associated as nicely. The equal not unusual experience is then carried out to cities, provinces, nations themselves, and plenty of others., to permit the neural network research new subjects approximately the world and replace its private records database. This is in aspect how Facebook's "People You May Know" feature works – figuring out the beginning of a courting regularly well-knownshows intimate records that could have lengthy beyond neglected even through the human beings involved. We do have such big statistics databases, but some thing like Wikipedia uses throngs of unpaid volunteers bitterly bickering for weeks approximately interpunction in difficult to understand articles to provide the majority of textual content utilized by the public; this type of a neural community might be used to both check a wiki or bring together a logo-new one.

Testing became finished with records databases, Claros and DBpedia. Though no longer the equal length, the neural network positioned the manner to efficaciously interpret items, records and individuals of the own family between them with a 99.Eight% accuracy. The authors then decided to up the mission with the aid of using casting off a random reality from every database and changing them with a diametrically opposite model of themselves, which means "guy" could have been modified through "female" and lots of others. This created a war we'd name a paradox, a statement that appears each real and fake on the same time, but the neural community managed to remedy on average 92% of all conflicts. The authors did examine that records supplied with a whole lot an awful lot less than 100% reality in the facts database did throw a wrench within the neural community's spokes.

Fooling the smart machine

Videos at the Open AI blog take a look at how picture classifying neural networks can be fooled with the resource of presenting a broadcast photo of a kitten digitally altered to contain blocky aftereffects. The photo is sincerely recognizable with the aid of the human eye, but a neural network sees a computer computer below nearly all angles and zoom factors, persisting despite the fact that the photograph is circled or moved apart.

The related 2018 studies paper titled "Synthesizing Robust Adversarial Examples" investigates the concept of turning 2D photographs and three-D-located gadgets right right into a supply of headaches for the neural network thru the usage of EOT (Expectation Over Transformation) set of hints that persists even if the image or object are grew to become round, filmed underneath brilliant lighting or hooked up zoomed in or out. In the instance tested in the paper, eight/10 photos of a three-d-revealed turtle have been recognized with the aid of the neural community as a rifle and the rest as "exceptional".

The 2015 research paper titled "DeepFool: a simple and correct method to idiot deep neural networks" describes an set of rules that provides minimal perturbations to any given picture to have the photograph reputation neural network see it as a few element else completely, an example examined being a whale diagnosed as a turtle. The DeepFool approach is then in comparison to similar perturbation algorithms in phrases of fee, tempo, and intrusion on the actual picture, discussing its use in statistics the structure of any given clever gadget and a manner to optimize the attack.

Chapter 6: Neural Network Algorithms

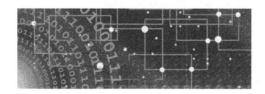

No doubt, you've heard the time period earlier than. It is often associated with all forms of technical mechanics but in latest years' algorithms are getting used in the development of automatic getting to know, the sector that is leading us to enhancements in artificial and computational intelligence.

To do that, the algorithms want to be flexible enough to evolve and make changes even as new statistics is obtainable. They are therefore able to deliver the desired answer whilst not having to create a selected code to remedy a problem. Instead of programming a rigid code into the device, the applicable records turns into part of the set of rules which in flip, allows the machine to create its private reasoning primarily based at the data supplied.

How does this art work?

This might in all likelihood sound a piece difficult however we'll attempt to interrupt this down into positive examples you could relate to. One of the 'getting to know' abilities of machines is the ability to categorise facts. To try this, the enter facts can be a aggregate of all kinds of records. The set of rules desires to pick out the different factors of the facts after which organization them into numerous one-of-a-kind classes based totally mostly on characteristics of similarities, versions, and other factors.

These tendencies may be any massive variety of factors beginning from figuring out handwriting samples to the types of files obtained. If this were code, the device need to handiest do one unmarried characteristic however due to the fact it's miles an set of policies which may be altered to wholesome a vast range of things, the computer can acquire this records and classify all forms of businesses that in form inside the specific parameters of the situations.

This is how machines can alternate their abilities to adapt to the state of affairs at hand. Your email account can observe all the emails you received, based totally totally on a sample which you have followed, and it divides them into first rate agencies. It can recognize which emails are important and you have to see right away, those which might be direct mail and junk mail, and even type out those that could pose a hazard to your computer as it contains a pandemic or malware.

With those styles of algorithms, machines can now research with the resource of looking your behavior and patterns and regulate their conduct consequently. So, the very thriller to a a fulfillment and effective neural pathway is based upon a wonderful deal at the algorithms your device uses.

At its maximum number one stage, device learning is using various preprogrammed algorithms that gather and take a look at information that permits you to decide feasible results inner the precise range. Each time those algorithms get keep of recent facts the device learns and adapts to beautify overall performance.

Commonly Used Algorithms You Should Know

Logistic Regression

Logistic regression might be very famous on the equal time as you want to clear up binary beauty problems. This is whilst the solution can be one in every of outstanding two options. Sometimes called dichotomy, it definitely works properly with problems that require both a real/faux or fine/no answer.

To understand logistic regression better, you first should have a clean information of linear regression. As an analyst, you need to discover the terrific line to show a selected fashion. This calls for locating an equation that offers the exceptional direct function that addresses the regression problem.

The stylish practice is to apply the least rectangular technique, wherein the concept is to shorten the space some of the road and the education statistics. Basically, the system is searching out a line that is "nearest" to all of the input data.

Logistic regression is part of a particular sort of set of guidelines called the generalized linear model. Unlike with linear regression, your aim is to discover a version that comes closest to the final value of the final effects or the variable. However, keep in mind that you are fixing a binary hassle so there may be no set fee to expect. It is just a depend of viable results. You're without a doubt seeking out the higher possibility that one final effects will truely occur.

Problems that linear regression ought to solve: How many inches of snowstorm can we get this yr?

Problems that logical regression can also remedy: Will it snow day after today?

Decision Trees

Decision tree algorithms are greater often used to classify a version and label it in a tree form. Many analysts find them to be superb gadget that would offer them correct and reliable output facts.

Decision bushes are easy to study and recognize. In truth, whilst using them you will be capable of see exactly why you want positive classifiers for you to make a choice. If you are new to writing code, this might be the tremendous set of policies to reduce your tooth on.

These algorithms all have exactly the same technique; to breakdown the information into the smallest viable subsets (those who encompass most effective one business enterprise of effects). The facts is cut up up based on some thing predictors are available. Then they group all subsets of the identical beauty together. They will maintain to try this till they have got the smallest set of information feasible.

Once that is finished, it's far very easy to make a prediction as to the expected conduct. Making this form of prediction is quite easy. All the device does is look at the path that suits the given predictors. It will result in the subset that contains all the sure answers.

Support Vector Machines

Support Vector Machines or SVM, are algorithms that may be used like guns.

This shape of set of rules is exceptionally complex and makes use of some of the maximum hard mathematical equations there are. Because of this specific complexity, SVMs can most effective slice via very small portions of datasets. So, if the initial education information is without a doubt too huge, it's far possibly that SVM isn't the pleasant choice.

Machine Learning and Big Data

Big Data is pretty masses what it looks like — the workout of coping with huge volumes of records. And by way of large, we're speaking approximately astoundingly huge portions of information — gigabytes, terabytes, petabytes of information. A petabyte, to location this era into attitude, is 10 to the 15TH bytes. Written out that is 1 PB = 1,000,000,000,000,000 byte. And those sizes are growing each day.

The term Big Data comes from the 1990s, no matter the truth that pc scientists have been handling big volumes of information for decades. What units Big Data other than records sets earlier than is the reality the size of facts units commenced to overwhelm the capability of conventional facts analytics software utility to deal with it. New database storage structures had to be created (Hadoop for instance) truly to maintain the statistics and new software program program software written in an effort to address a lot records in a vast way.

Volume

The time period quantity refers back to the big amount of statistics available. When the time period Big Data modified into coined inside the early 2000s, the amount of records to be had for assessment modified into overwhelming. Since then, the quantity of statistics created has grown exponentially. In reality, the amount of statistics produced has emerge as so large then new storage solutions had to be created just to address it. This growth in available information suggests no signal of slowing and is, in fact, growing geometrically via doubling each years.

Velocity

Along with the rise in the quantity of statistics being created is the charge at which it is produced. Things like smartphones, RFID chips, and real-time facial recognition produce now not best massive quantities of facts, this data is produced in actual time and should be dealt with as it is created. If now not processed in actual time, it should be stored for later processing. The growing tempo of this information arriving lines the capability of bandwidth, processing strength, and garage region to incorporate it for later use.

Variety

Data does no longer get produced in a unmarried format. It is saved numerically particularly databases, produced in shape-an awful lot much less text and e-mail files, and saved digitally in streaming audio and video. There is stock marketplace statistics, monetary transactions, and so forth, it all uniquely dependent. So now not most effective have to large quantities of records be dealt with proper away, it's miles produced in lots of formats that require outstanding techniques of dealing with for each type.

Lately, more V's have been brought:

Value

Data is intrinsically valuable, but best if you are capable of extract this fee from it. Also, the nation of input facts, whether or not or no longer it's miles nicely established in a numeric database or unstructured text message chains, affects its rate. The a lot an awful lot less shape a facts set has, the greater paintings wants to be placed into it in advance than it could be processed. In this sense, well-based totally completely information is greater valuable than tons much less-mounted facts.

Veracity

Not all captured information is of same fantastic. When managing assumptions and predictions parsed out of huge statistics sets, understanding the veracity of the information being used has an crucial impact on the weight given to the facts analyzing it generates. There are many motives that restriction statistics veracity. Data may be biased with the beneficial useful resource of the assumptions made with the aid of the use of folks who accrued it. Software insects can introduce errors and omission in a information set. Abnormalities can reduce records veracity like at the same time as wind tempo sensors subsequent to each distinct file super wind pointers. One of the sensors can be failing, however there may be no way to decide this from records itself. Sources can also be of questionable veracity — in a commercial enterprise company's social media feed are a chain of very terrible opinions. Were they human or bot created? Human mistakes, as in a person signing as an awful lot as an internet company enters in their telephone quantity incorrectly. And there are various more techniques facts veracity may be compromised.

The thing of dealing with all this information is to become aware about beneficial element out of all of the noise — corporations can locate techniques to reduce costs, increase pace and efficiency, layout new products and types, and make extra smart alternatives. Governments can find out similar blessings in studying the records produced with the resource of using their residents and industries.

Here are a few examples of modern-day uses of Big Data.

Product Development

Big Data can be used to predict consumer demand. Using cutting-edge-day and beyond products and services to classify key attributes, they might then model the ones attributes' relationships and their fulfillment within the marketplace.

Predictive Maintenance

Buried in based completely records are indices that might count on mechanical failure of gadget additives and systems. Year of manufacture, make and version, and so forth, offer a manner to are expecting destiny breakdowns. Also, there may be a wealth of unstructured data in mistakes messages, provider logs, running temperature, and sensor records. This statistics, at the same time as correctly analyzed, can count on problems earlier than they take location so safety can be deployed preemptively, reducing every fee and tool downtime.

Customer Experience

Many organizations are not anything with out their clients. Yet obtaining and keeping customers in a aggressive landscape is tough and high-priced. Anything that might deliver a business business business enterprise a place may be eagerly implemented. Using Big Data, agencies can get a miles clearer view of the consumer revel in via examining social media, internet website on-line visit metrics, call logs, and some different recorded consumer interaction to alter and beautify the consumer revel in. All inside the pursuits of maximizing the fee brought in an effort to collect and hold clients. Offers to person customers can emerge as no longer exceptional greater personalized however greater applicable and accurate. By using Big Data to discover complicated troubles, corporations can control them brief and effectively, reducing consumer churn and awful press.

Fraud & Compliance

While there can be unmarried rogue horrible actors to be had in the virtual universe searching for to crack device safety, the real threats are from organized, properly-financed groups of professionals, now and again agencies supported with the beneficial aid of foreign governments. At the equal time, protection practices and requirements by no means stand despite the fact that but are continuously converting with new technology and new techniques to hacking present ones. Big Data helps select out records styles suggesting fraud or tampering and aggregation of those massive statistics devices makes regulatory reporting lots quicker.

Operation Efficiency

Not the sexiest difficulty be counted, but that is the location wherein Big Data is currently providing the most price and return. Analyze and test out manufacturing structures, look at consumer remarks and product returns, and have a look at a myriad of different organization elements to lessen outages and waste, or maybe expect future call for and trends. Big Data is even useful in assessing modern-day desire-making techniques and the way well they function in assembly call for.

Innovation

Big Data is all approximately own family individuals among tremendous labels. For a large company, this can suggest examining how human beings, institutions, extraordinary entities, and agency techniques intersect, and use any interdependencies to electricity new strategies to take gain of those insights. New tendencies may be expected and present trends can be higher understood. This all results in understanding what clients really want and expect what they may need in the destiny. Knowing sufficient approximately person customers can also cause the capacity to take benefit of dynamic pricing models. Innovation driven by manner of Big Data is actually only confined with the aid of using the ingenuity and creativity of the humans curating it.

Machine Learning is also supposed to deal with huge portions of records proper away. But on the same time as Big Data is centered on using present information to discover developments, outliers, and anomalies, Machine Learning makes use of this equal records to "observe" those styles with a view to deal with future statistics proactively. While Big Data appears to the beyond and gift information, Machine Learning examines the prevailing facts to discover ways to address the information as a manner to be gathered inside the destiny. In Big Data, it's miles folks that outline what to look for and a manner to put together and form this records. In Machine Learning, the algorithm teaches itself what is important thru new release over test statistics, and while this gadget is finished, the set of policies can then bypass beforehand to new information it has in no manner professional earlier than.

What is Predictive Analytics?

For severa groups, large information - genuinely large volumes of unstructured, semi-primarily based and uncooked established facts is an untapped deliver of facts that could useful resource enterprise selections and improve operations. As this data maintains to alternate and diversify, more and more corporations are taking to predictive analytics to faucet the deliver and make benefits from the big-scale facts.

There is a commonplace miscomprehension that system mastering and predictive analytics are one and the equal. That isn't the case. They do overlap in a single area however and this is predictive modeling. Basically, predictive analytics includes a number of statistical techniques which embody ML, facts mining and predictive modeling and uses historical and modern-day information for estimating or predicting destiny very last outcomes. This very last consequences might be the behavior of a client in some unspecified time in the destiny of the acquisition or probably changes within the market. It permits us to guess the viable future occurrences with the assessment of past sample.

How Does Predictive Analytics Work?

The predictive analytics gets driven with the resource of the use of predictive modeling. It is an technique rather than a technique. ML and predictive analytics are hand-in-hand due to the truth the predictive models commonly encompass ML algorithms. The created models may be expert over a time body to react to new values or special statistics thereby turning inside the outcomes wanted via the commercial organisation commercial enterprise organization.

There are varieties of predictive fashions. One is the magnificence version which predicts elegance membership and the second one is the regression version that predicts numbers. The models are made from algorithms which carry out facts mining and statistical evaluation to determine patterns and developments inside the information. The predictive analytics software program application may also moreover have integrated algorithms which may be used to create predictive fashions. Algorithms are known as as classifiers and that they select out the set of categories to which the records belongs.

Commonly Used Predictive Models

The most extensively used predictive models are:

- Regression (Linear and logistic)
- Decision Trees
- Neural Networks

Regression (Linear and logistic)

It is one of the greater famous techniques available in statistics. Regression evaluation affords a relationship between variables and exhibits key patterns in various and massive information devices. It moreover exhibits out how they relate to every extraordinary.

Decision Trees

The decision trees are easy but powerful varieties of multiple variable evaluation. Decision bushes are produced with the aid of the use of algorithms which perceive one of a kind methods of splitting the facts into branch-like segments. They partition the data into subsets depending on numerous classes of enter variables. It allows you recognize a person's path to a choice.

Neural Networks

They are constructed at the patterns of the neurons in the human mind. Neural networks are often known as artificial neural networks and are a variance of deep learning generation. They are usually used to resolve hard sample recognition conditions and are unbelievably useful for reading large records devices. They are terrific at handling nonlinear data relationships and additionally paintings properly whilst some variables are unknown.

Classifiers

Every classifier strategies the information in a notable manner so, for the managers to get the results they require they want to pick out the right classifiers and fashions.

• Clustering algorithms: They put together information into particular companies with similar people.

• Time Series algorithms: They plot the information sequentially and are beneficial in forecasting regular values over a time body.

• Outlier Detection algorithms: They recognition surely on anomaly detection, identifying events, observations or devices that don't comply with a specific predicted sample or requirements in a statistics set.

• Ensemble Models: These models appoint severa ML algorithms for acquiring a higher predictive overall performance than compared to the output anticipated from a single set of policies.

• Naive Bayes: This classifier allows you to are looking forward to a class or a class primarily based on a provided set of skills with the useful resource of the usage of opportunity.

● Factor Analysis: It is a manner used for describing versions and goals at locating independent latency in variables.

● Support Vector Machines: They are a supervised form of device gaining knowledge of approach that makes use of associated gaining knowledge of algorithms for reading facts and recognizing patterns.

Machine Learning and Predictive Analytics Applications

The organizations which can be overflowing with information are suffering to expose all the statistics into useful belief. For these organizations, ML and predictive analytics can set up the solution. No recollect how massive the records is if it can't be used to enhance the outside and internal techniques and meet goals it becomes a useless aid. Predictive assessment is used extra usually in advertising and marketing, protection, risks, fraud detection and operations. Here are a number of the examples of the manner device learning and predictive analytics are utilized in numerous industries:

Financial Services and Banking: In the economic offerings and banking employer, ML and predictive analytics are used collectively to measure market risks, encounter and reduce fraud, understand opportunities and there are numerous special makes use of.

Security: Cyber safety is on the top of the time desk for nearly all corporations in the present day international. It is not any marvel that ML and predictive analytics play a key position in safety elements. The safety companies use predictive evaluation frequently to enhance their everyday performance and services. They can come across anomalies, recognize client behavior, come upon fraud and as a end result, they beautify information protection.

Retail: The retail enterprise is the use of ML for knowledge consumer behavior higher. Who is shopping for what and in which? They need to apprehend the solution to those queries. These questions can be spoke back with accurate predictive models and statistics gadgets thereby supporting shops to plan earlier and inventory devices based on client traits and seasonality. Improves the ROI a first rate deal.

Developing The Right Environment

Although predictive analytics and ML can be a big improve for maximum businesses, enforcing those solutions halfheartedly without interest for his or her fitment into regular operations will simplest prevent their potency to deliver the perception the employer desires. To get the high-quality out of ML and predictive analytics, organizations want to make certain that they've the structure to assist the answers on the component of exquisite records so you can help them in reading. Data schooling and its extraordinary are the vital difficulty portions of predictive analytics.

The input records which might also moreover span during severa structures and encompass more than one facts sources need to be centralized and unified in a coherent manner. For undertaking this the businesses want to increase accurate dependable records governance packages to govern the general records manage and make sure that most effective the brilliant data gets captured and used. Another aspect is, the current-day tactics may need to be altered to embody ML and predictive analytics as this could permit groups to have usual performance in any respect factors of the economic employer. And most importantly the groups need to apprehend what troubles they want to be resolved as it will useful resource them in figuring out the maximum suitable version for use.

Predictive Models

The IT professionals and facts scientists operating in an business enterprise are commonly tasked with choosing or growing the proper predictive models or likely gather one for themselves to satisfy the company desires. However, nowadays the ML and predictive analytics isn't virtually the location of expertise for mathematicians, statistics scientists, and statisticians but there are company professionals and analyst strolling in the location. More and additional people in corporations are the usage of the fashions to increase insights and decorate operations. However, there are problems after they lease are not aware of what model to use or the manner to install it or inside the case after they want some facts right now. There is cutting-edge software application available to help the employees with the problem.

What is Data Mining?

Data mining manner extracting knowledge out of huge quantities of information. In one-of-a-type words, we are capable to say that it's far a method of discovering particular sorts of styles inherited inside the statistics units which might be new, beneficial and accurate. Data mining is an iterative gadget which creates descriptive and predictive models thru uncovering formerly unknown styles and tendencies in a massive amount of records. This exercising is accomplished to help preference making. It is essentially a subset of corporation analytics and is just like experimental studies. Origins of information mining may be positioned in data and databases. ML, rather, works with algorithms that enhance mechanically through enjoy they benefit out of statistics. In precise phrases, in machine analyzing, we find out new algorithms from revel in. These algorithms of ML can extract facts mechanically however the deliver used for device studying is likewise facts. It entails types of statistics, one is test information and the second is the schooling statistics. Data mining strategies are generally utilized in tool gaining knowledge of and along side the gaining knowledge of algorithms it's far used to assemble models of what's taking place backstage to predict the final effects of the destiny.

What is information mining and what's the connection amongst ML and information mining? Data Mining way extracting expertise out of a large quantity of facts. It turned into added in 1930 and before everything it emerge as called facts discovery in database. Data mining is carried out to get regulations out of present statistics. Its origins lie in traditional databases having unstructured records. It is carried out wherein you can broaden your own fashions and information mining techniques are used. It is greater natural and involves greater involvement of humans. They are applied in cluster assessment. Data mining is abstracted from records warehousing. It is more of a research the use of strategies much like ML but is applied in constrained sectors.

Data Mining Techniques

The experts walking within the location of information mining rely on strategies and intersection of facts, database manage, and tool learning. They have devoted their careers to facts what conclusions are to be drawn from a huge quantity of facts. What are the strategies used for turning this into fact? Data mining is effective whilst it draws on some of those strategies for their assessment:

1. Tracking Pattern

One of the fundamental techniques implemented in facts mining is studying to understand styles in the data gadgets. Normally that is an aberration within the facts which is going on at a few intervals or a flaw or an ebb in some variables over a term. For example, you can have a study that sale of superb product spike up straight away in advance than the vacations. Or you may word that heat climate drives human beings on your internet internet page.

2. Classification

It is a extra complex information mining technique that asks you to accumulate one in all a type attributes collectively in discernable lessons which can be later used to attain at further conclusions or serve in some other feature. For instance, if you are evaluating the information on independent purchaser's economic background and purchase statistics you'll be capable of classify the people as immoderate, medium or low-threat applicants for credit score score. You can then use the classifications to investigate more approximately the customers.

three. Association

This is associated extra to monitoring patterns however, it's miles more specially concerned within the dependently associated variables. In the case of an affiliation, you look for particular attributes or occasions which can be correlated to special attributes or events.

For example, you can have a look at that after your patron purchases a few unique item similarly they purchase every other related item. This series of events is used to populate the "humans also presented" phase inside the on line keep.

4. Outlier Detection

In some instances absolutely identifying the overreaching sample cannot provide you with a smooth expertise of the statistics set. You are also required to recognize the anomalies additionally known as outliers within the statistics. For example, your customers are almost completely male, however, in some unspecified time inside the future of a single week in July there can be a shocking rise in lady clients. You may furthermore want to research the cause for the event and discover what drove the profits so, you can both reflect it or recognize the conduct of your goal market higher.

5. Clustering

This is much like kind, but, includes grouping of chunks of facts collectively which is similar. For example, you may select to cluster precise demographics of the customers in numerous packets based totally on how lots greater income they earn or how regularly they may be shopping for on the internet keep.

6. Regression

Regression is used basically as a form of modeling and making plans. It is used to find out the chances of presence of positive variables due to the reality a few other variables are there. For example, you may use this to mission a few charge based totally totally on factors which includes consumer call for, competition, and availability. More specifically the principle interest of regression is in helping you discover particular relationships amongst or extra variables inner a specific data set.

7. Prediction

It is without issues one of the most precious data mining techniques used. This is due to the truth it's miles used for predicting the kind of information you could see within the destiny. In some cases, simply with the useful resource of understanding and spotting the ancient inclinations we are able to chart an correct prediction of what goes to reveal up within the future. For example, we are able to see the credit history of clients and their past purchases to anticipate whether or not there could be a credit rating danger in destiny in case a mortgage is extended.

Recurrent Neural Networks

In this bankruptcy, we are going to flow directly to every different cutting-edge-day, novel neural network structure called the recurrent neural network.

Whereas convolutional neural networks allow us to paintings with uncooked snap shots, recurrent neural networks allow us to paintings with raw sequences.

As with pics, you may nevertheless "make" a series paintings with a ordinary feedforward neural network, however this is neither green nor effective.

Recall that CNNs will assist you to have fewer parameters than an ANN. This is authentic with RNNs as properly.

As promised, we're nonetheless not going to get into the math on this ebook, so you'll clearly need to receive as real with me in this.

Since the priority of this ebook is prepared know-how the Keras API vs. The underlying math, we are going to bypass over the equations concerned in an RNN.

Let's begin through thinking about what the facts want to look like.

What is a series?

It's a list of ordered devices. E.G. 1, 2, three, four, 5, ...

One instance of a chain is the USD-EUR trade rate through the years.

One may also moreover don't forget a series like this as a time-collection or greater commonly a "signal".

Note that an photograph is likewise a sign, but it's far a 2-D sign (the two dimensions are top and width). A time-series is a 1-D signal due to the fact there's best 1 dimension (time).

As a side check, you would possibly then surprise: if CNNs manner indicators and RNNs approach sequences, however sequences are without a doubt surely 1-dimensional indicators, can CNNs be used for sequences and RNNs be used for pics? The solution is positive! Researchers have attempted this and decided this approach to obtain success.

In reality, CNNs are in the decrease returned of cutting-edge-day speech era systems like Google's WaveNet.

But we are becoming a touch in advance of ourselves right here.

What is the most effective time collection you could remember?

How approximately a sine wave?

Can we build an RNN to are expecting the following value on this time-series, given a difficult and rapid of previous values?

We excellent can!

Chapter 7: The Neuron

As mentioned, neurons are the most essential constructing blocks of our mind and billions of them are associated with execute our idea methods. Their connections and systems are constantly converting, allowing us to analyze and adapt to our surroundings. Let's check the shape of this first rate cell.

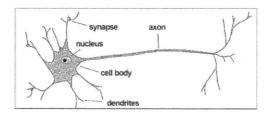

As you can inform, neurons have a primary body and branching units of connections: the enter structure (dendrites) and the output form (the axon). The axons connect to the dendrites of various neurons through the synapse, ultimately forming a connection.

Signals are transmitted amongst neurons within the form of electro-chemical indicators. A neuron first receives a sign thru the dendrite to the mobile body – the signal is processed, altered and assessed by using using the nucleus. However, the neuron may be activated fine if the input symptoms exceed a positive quantity inner a quick period (i.E. The facet). If this quantity is reached, the neuron will become activated and it fires a sign to the connected neurons through the axons. This is how neurons are interconnected and alternate facts.

Artificial Neurons

Now that we've got were given tested the simple shape of a neuron, we will have a look at the way to model its features and operation using software program software. We create synthetic neurons as mainly clean mathematical features – their output is calculated the usage of three critical factors:

• Input signs: the ones originate right now from neighboring neurons.

• Weights: every neuron applies a completely specific weighting difficulty to each enter sign acquired.

• Bias: this rate is precise to the neuron and is executed to the general output.

Notice that the output isn't the easy sum of all of the weights, it should be prolonged with the useful resource of way of an activation characteristic. Most synthetic neurons are modeled the usage of a sigmoid activation function, as validated beneath. This is a smooth, non-save you and generally growing curve (i.E. Gradient is normally incredible). The mathematical equation of the sigmoid function is $f(x) = 1/(1 + e-x)$.

Perceptron

In chapter 7 we mentioned how synthetic neurons emit signs consistent with the sigmoid activation characteristic based totally mostly on three parameters: enter signals, enter signal weight and bias. In this bankruptcy we can use this recognize-a manner to construct a completely easy neural network – known as a perceptron. In fact, this community is so easy it simplest contains one output neuron! Nevertheless, it includes the essential issue behaviors determined in nowadays's neural networks.

In precise, perceptrons display how artificial neurons manner facts and how neural networks optimize the weights in every connection to look at. Due to their simplicity, perceptrons have very limited applications and can most effective cope with especially easy issues. I actually have recommended the schematics of a easy perceptron on the subsequent net web page.

How Do Perceptrons Learn?

Now we are becoming to the exciting bits! Let's speak how a perceptron learns and adapts to resolve a hassle. Please recognize we're dealing with the equal of a unmarried 'brain cell' – the issues it can clear up is therefore very restrained.

Perceptrons have a look at the usage of supervised training, meaning you need to provide a set of inputs and their matching output. The perceptron works thru the input and computes an output – if it isn't always correct the weights are adjusted until the right solution is acquired.

Learning is as a result an iterative way, this means that man or woman weights must be tuned and changed over and over to discover the maximum suitable combination. To find out the maximum appropriate weights you want to use the subsequent formula:

wtnew = wtold + α(preferred – output)*input

Where wtnew and wtold are the weights of a single connection, α is the getting to know fee (i.E. How rapid a neural community adapts), favored is the precise output, output is the modern-day output and enter is the input rate for the contemporary-day connection.

This method also can moreover seem a piece perplexing, so allow's dive into an instance – this could make the whole lot less complicated to recognize. Below I clearly have shown a totally skilled perceptron (i.E. Weights are already assigned) and the output is tested.

In the perceptron you may word three inputs (1,zero,1), 3 weights (zero.Five,0.2,0.Eight) and an output (1). Now, I need to find out the weights a good way to produce the output fee zero, how can I do that? You must observe the previous equation for each weight. Take a look below at my calculations (I truly have assumed $\alpha = 1$ for simplicity).

Top weight:

wtnew = wtold + α(preferred − output)*enter

wtnew = zero.Five + 1*(zero − 1)*1 = zero.Five

Middle weight:

wtnew = wtold + α(preferred − output)*enter

wtnew = zero.2 + 1*(zero − 1)*0 = 0.2

Bottom weight:

wtnew = wtold + α(preferred − output)*input

wtnew = 0.8 + 1*(0 − 1)*1 = -zero.2

We have calculated new weights for the perceptron, which may be cited inside the following internet web page. You will notice the contemporary and updated weights are delivered. However, getting to know is an iterative method and the weights can however be tuned. As a end result, we ought to use the activation function to calculate the cutting-edge output and decide if it is high-quality (i.E. Is it near enough to our favored output?). If it is not close to enough, we have

149

to repeat the machine above till the current output fits our favored output. This is the concept of getting to know in neural networks.

Multilayer Feedforward Networks

Multilayer Feedforward Networks (MFN) are with the beneficial useful resource of a long way the most commonplace and substantially used Neural Network structure. They are an instantaneous extension of the perceptrons we studied in the previous bankruptcy and, basically, art work inside the identical manner.

However, MFNs are used to clear up complicated issues requiring good sized computational power and cognitive talents – those can not be carried out the usage of a single neuron. In fact, MFNs are constructed using severa artificial neurons, which may be organized in layers.

1. Input layer: those layers introduce input statistics and starting conditions into the network.

2. Hidden layer: the ones neurons generally perform elegance primarily based on precise and statistical capabilities within the data. Two hidden layers are generally enough to remedy maximum troubles, without requiring immoderate complexity or computationally-heavy optimization strategies.

3. Output layer: the end end result of all computations inside the community are presented to the outside world via the ones neurons.

The key functions of this architecture are:

• Information best travels in a unmarried direction (enters at input, via the hidden layer and output consequences are high-quality emitted with the useful resource of the output layer)

• Neurons in the equal layer are not linked to every awesome

• Hidden layers may be brought or removed to effortlessly regulate computation complexity whilst no longer having a main effect on the general network

Below you may view a totally important instance of a MFP with 3 layers and five neurons. You can see the neurons inside the enter layer denoted with the aid of using the letter I, the neurons in the hidden layer are cited the usage of the letter H and the output layer neurons are stated the use of the letter O. In this network, I actually have moreover proven examples of all weights and biases, as you'll find out in a actual MFP.

Please recognize that may be a easy network and can incredible be used to remedy simple, essential calculations. To give you a higher understanding of what's used by the corporation, Google uses a contemporary MFP in its internet are seeking algorithms — this has nine hidden layers!

Backpropagation

When first building a neural community, you are arranging 'blank' neurons – those have random weights and biases connected to them. Clearly, this new community can't be used to treatment problems. However, in case you educate it with training facts and practice mastering algorithms (for example the mathematical expression used in financial disaster 8 for the perceptron), the community will learn how to remedy the problem. You do now not want to go into new formula, reprogram any code or perform direction & errors parameter tuning; the Neural Network will adapt its functionality on your hassle. Training a community to resolve a hassle requires minimum potential and try, exceptional sufficiently big education information gadgets and time.

Backpropagation is the most not unusual set of regulations Neural Networks use to take a look at. It is a important aspect of all modern-day networks and vital to their significant and clean implementation. This set of regulations is based mostly on supervised training (for a proof of supervised vs unsupervised education go to financial spoil five). Meaning, which will teach a network to remedy a hassle you ought to have a hard and fast of inputs and outputs (i.E. Beginning conditions and the corresponding answers).

Let's take a step-through-step have a have a look at Backpropagation and how it lets in Neural Networks to research:

Step 1 – User collects education statistics.

Before schooling the community, you need to first discover a difficult and rapid of instance troubles. The community will examine those troubles and apprehend precisely the way to remedy the problem. Since Backpropagation is a supervised reading set of regulations, your examples need to incorporate the enter facts and the tremendous output solution.

Step 2 – Feed input records into the network through the input layer

You need to feed the input facts of all of your examples into the enter layer of your community.

Step 3 – Network techniques information to supply an output

The enter statistics is processed with the resource of using the enter layer and via the hidden layer. As the signal passes through those layer, every neuron applies capabilities based person weights and biases. The prevent end result is generated by way of way of the use of the final neurons, determined within the output layer.

Step 4 – Calculate the mistake rate

For each set of enter, the Backpropagation set of guidelines compares the actual output and the expected output. If those values do not in shape, there must were a computational errors within the network – this may be quantified using the error feature. The maximum common errors function used is the endorse squares errors.

Step five – Update neuron weights/biases & iterate!

Once the error price has been calculated, the weights and biases of each unmarried neurons are regularly changed. First the weights and biases inside the output layer are first-rate-tuned, on foot backwards into the hidden layers and eventually into the enter layer. This is why the learning set of pointers is referred to as 'Back'-'propagation' – tuning starts offevolved in the very last layer and little by little propagates once more into the network.

With every new weight and bias, the network's calculations are repeated and the present day errors is tested. Throughout those kinds of iterations Backpropagation has best one purpose: to reduce the mistake feature.

Finally – the Network has Learned!

Alternative Architectures

Thus a protracted way we have have been given first-class covered forms of neural internet structures: the perceptron (maximum number one) and multilayer feedforward networks (the maximum famous). Many unique styles of neural networks exist and on this bankruptcy I want to provide you with a short publicity to unique famous Neural Network architectures.

Counterpropagation (CP) Networks

The most substantial benefit of this community lies in mastering tempo: schooling a modern-day day CP network can be about 100 instances faster than a everyday multilayer feedforward community. However, as soon as informed CP networks show off substantially worse generalization – which means that that they war to cope with new and unseen enter information, even though it's similar to the education statistics. In terms of form, CP networks simplest consist of three layers of artificial neurons:

• Input layer: as in preceding systems, the neurons on this network first technique the raw input facts.

• Kohonen Layer: all neurons on this layer take a simple weighted sum of their inputs. The neuron with the most vital weighted sum emits a 1 – all the exceptional neurons emit 0.

• Grossberg layer: Each neuron inside the Grossberg layer can get hold of quality 1 enter sign (from Kohonen layer), it's then used to calculate the final output of the community.

Many don't forget that CP networks are a higher example of the human mind than ordinary multilayer feedforward networks. This is because every layer in CP community have end up designed with a unique and outstanding reason. The Kohonen splits input statistics into separate classes and the Grossberg without difficulty controls the network's very last output the usage of its weights (due to the truth that those neurons handiest have 1 enter sign).

Recurrent Neural Networks

Recurrent Neural Networks (RNN) are very unique – they preserve previous calculations and use them to beautify the accuracy of destiny calculations. In reality, the outcomes produced via the output layer are fed yet again proper away into the enter layer of the network.

RNN can keep statistics about time and therefore are most excellent to forecasting applications, i.E. Figuring out styles in the stock marketplace or predicting the climate.

Self-Organizing networks

Self-organizing networks use unsupervised studying and consequently may be used with unlabeled information units.

These maps best have 2 layers of neurons: the input and output layer and are essentially created 'smooth'; on the same time due to the fact the connections are installation at a later diploma. In fact, finally of training the community analyzes the enter statistics and searches for statistical similarities. Based on those statistical similarities, output neurons emerge as connected to enter statistics. In essence, this network represents a totally advanced clustering algorithm.

Chapter 8: Predicting Time Series

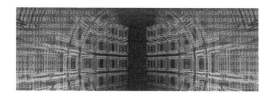

The relevant report for this example is sine.Py.

Warning: in case you're now not coding on the aspect of me thru now, this monetary smash goes to get REALLY complicated.

Let's begin with the imports.

import numpy as np

import pandas as pd

import matplotlib.Pyplot as plt

from keras.Fashions import Model

from keras.Layers import Input, SimpleRNN, Dense

The most effective new layer we want here is the SimpleRNN.

The first hassle we need to do is generate the facts.

make the particular statistics

series = np.Sin(zero.1*np.Arange(hundred)) + np.Random.Randn(two hundred)*0.1

plot it

plt.Plot(series)

plt.Display()

We've moreover introduced some noise (due to the fact "actual" data is noisy), and plotted it so that you understand what the statistics looks as if (no longer which you shouldn't understand what a loud sine wave seems like already).

Next, we want to build our dataset.

This calls for a chunk thinking about what a sequential dataset have to appear like.

Recall that for feedforward ANNs, a records matrix X is a 2-D array of form NxD.

For a CNN, a records matrix X is a 4-D array of form NxHxWxC.

For an RNN, a data matrix X is a three-D array of form NxTxD.

As regular, N refers back to the type of samples.

T refers back to the series duration (you can do not forget this with the aid of way of noting that the phrase Time starts with the letter T).

D (as inside the case of ANNs) represents the vector dimensionality.

Thus, an RNN is "like" an ANN, in which the enter is a function vector of duration D, besides that for each of the N samples, we consider T of these D-sized characteristic vectors in a sequence.

In the case of a sine wave, D=1 on the grounds that there is best one dimension.

What is T?

In this situation, we're allowed to pick out out T.

We've generated a sine wave of period two hundred.

You can take into account decreasing the sine wave into overlapping domestic windows, every of period T.

Let us count on T=10.

That approach each pattern collection X[n] will encompass 10 sequential values from the sine wave, and Y[n] (the corresponding goal) may be the eleventh charge of the sine wave.

In exclusive phrases, our RNN version will learn how to soak up as enter a sequence of 10 steps in a time collection, and forecast the subsequent step.

Let us unique that in code:

T = 10

D = 1

X = []

Y = []

for t in variety(len(series) - T - 1):

```
x = series[t:t+T]

X.Append(x)

y = series[t+T]

Y.Append(y)

X = np.Array(X)

Y = np.Array(Y)

N = len(X)
```

We begin thru initializing X and Y to empty lists.

Since every pattern will consist of T+1 values (T for the enter and 1 for the output) we're able to first-rate loop as an awful lot as len(series) - T - 1.

The T samples similar to the input are series[t:t+T] and the 1 pattern similar to the output is series[t+T].

We append the ones to X and Y at every new launch of the loop.

After the loop is entire, we cast X and Y to Numpy arrays.

Note that at this element, X is exceptional a 2-D array of form NxT (that is because of the fact every x we introduced to X changed right right into a 1-D array of length T).

Thus, so as for X to be in an appropriate layout for the RNN (a three-D array of form NxTxD) we want to feature the more 1 dimension to make it NxTx1.

inputs = np.Expand_dims(X, -1)

Next, we construct the version:

i = Input(form=(T, D))

x = SimpleRNN(five)(i)

x = Dense(1)(x)

version = Model(i, x)

Much simpler than a CNN and nearly as easy as an ANN. Nice!

As famous, we begin with an Input layer. As anticipated, we specify the input form as (T, D) (keep in mind the sort of samples N is implicit).

We bypass i via a SimpleRNN layer.

166

We create a very last Dense layer to map the SimpleRNN's hidden values to a unmarried rate (as a give up result the 1 argument within the Dense constructor).

Unlike the previous examples in this e-book, we are now acting regression in desire to class.

Since we're doing regression, we don't want the final dense layer to have any activation feature the least bit. Recall that the softmax activation bounds the output to be among 0 and 1, in order that it represents a legitimate opportunity distribution.

Regression outputs can address any price due to the reality you're no longer trying to count on a class, you're seeking to are anticipating some of.

model.Collect(

 loss='mse',

 optimizer='adam',

Accordingly, we additionally don't want to apply the categorical_crossentropy loss, because of the fact that's for type. Instead, we need the propose squared error, targeted as "mse".

As elegant, we use the "adam" optimizer.

Next, we healthy the facts:

```
r = model.Match(

inputs[:-N//2], Y[:-N//2],

batch_size=32,

epochs=eighty,

validation_data=(inputs[-N//2:], Y[-N//2:]),
```

I've determined at the number one 1/2 of of the records to be the training samples and the second half of of to be the validation samples. Why?

Unlike with class, wherein we are able to shuffle the statistics, this doesn't sincerely make experience on the same time as you're managing time series.

For type, it makes experience to shuffle your statistics as it's now not time-based totally, a picture of a cat looks as if a photo of a cat no matter while it become taken. A picture of a cat from the destiny will probable look similar.

For time series, this idea of the "future" certainly topics. I don't want to combine information. Suppose I in reality have information from days 1, 2, three, …, 10.

If I educate on days 2, four, 6, 8, 10, and check on days 1, 3, 5, 7, 9, then I'm now not honestly predicting the destiny. I'm the use of the destiny to anticipate the past. That's not beneficial.

What is useful is schooling on days 1, 2, three, four, 5 and the use of that model to expect days 6…10.

Of route, in our smooth sine wave example this doesn't truely rely quantity due to the fact a sine wave simply repeats the equal pattern time and again over again. For actual time-series, it does be counted number.

Next, as everyday, we plot the loss:

```
plt.Plot(r.Records['loss'], label='loss')
```

```
plt.Plot(r.Records['val_loss'], label='val_loss')

plt.Legend()

plt.Display()
```

Note that for regression problems, there may be no perception of accuracy. If the goal is five, and you predict five.000001, intuitively, you understand that your prediction is close to the aim. That is a first-rate prediction. But five != five.000001, so it's now not "correct". Thus, any concept of accurate charge or mistakes fee proper here is beside the point.

In a kind script, things might likely prevent proper right here due to the fact we've positioned that the loss has converged to almost 0.

In regression, for the purpose that that is a time-series, we are succesful to plan our predictions inside the direction of the proper objectives!

```
outputs = version.Predict(inputs)

print(outputs.Shape)

predictions = outputs[:,0]
```

```
plt.Plot(Y, label='dreams')

plt.Plot(predictions, label='predictions')

plt.Title("many-to-one RNN")

plt.Legend()

plt.Show()
```

Note that for the reason that final layer is a Dense(1), the output back with the aid of model.Expect() is of shape Nx1.

In order to plan this information, we pick the primary (and simplest) column the use of outputs[:,0].

Neural Networks And Artificial Intelligence

Only human beings have the ability to assume, examine, and understand their surroundings, its moves, and the realities that exist. Only humans can give an explanation for their worldwide and think in the summary. Our brains can absorb a wealth of information and draw conclusions in the fraction of a 2d.

Machines, as a substitute, the use of the identical set of facts, will face numerous traumatic conditions. For instance, these days, computer systems can maintain a database of hundreds of comparable pictures, check them and draw a conclusion with 90 5% performance. This is pretty excellent by using the use of way of any stretch but despite the fact that, it can't deliver an reason for why it selected the images, verify their because of this, or distinguish why one image isn't much like any other. In other words, pc systems can compute however they are capable of't purpose. So, despite the reality that they will be capable of producing fantastic results with the duties they may be given, they nevertheless are a long manner in the returned of the capacity of the human thoughts in masses of approaches.

To triumph over those boundaries, a totally specific shape of machine getting to know has been superior. Most parents are aware about it as Artificial Intelligence or AI. The time period refers back to the simulation of intelligence in machines. Today's machines are programmed to 'count on' and mimic the way the human thoughts operates. These machines are slowly taught to rationalize in given conditions, have a look at, and pick out out out a path of movement that would have the great hazard of conducting the remaining reason.

As technology keeps to expand, it is quality natural that older machines will become vintage. For example, machines that perform number one functions or can recognize and perceive text became reducing aspect however these days, they are so out of date that they are now not taken into consideration to be synthetic intelligence. This is due to the fact the characteristic now could be taken with no consideration. This ability has emerge as so not unusual that it is regularly taken into consideration as a ordinary computer function.

Computers stated to have artificial intelligence these days are people who show greater top notch competencies like being able to play chess, self-driving vehicles, and clever houses. Why are the ones features given the label of artificial intelligence? For those that could play chess, the ones machines can win the game in opposition to a human opponent, self-using automobiles have mastered the ability to take in all the out of doors statistics surrounding them and compute it rapid enough that they are capable of navigate and attain their vacation spot without inflicting an accident of some kind. Smart homes can have interaction with participants of the circle of relatives, manage the temperature, control safety, or even food supply to make sure the comfort of its population.

Of route, on the identical time as anything surely new is brought to the arena, there may be regularly some skepticism involved and the same may be actual with regards to artificial intelligence. Even even though we engage with AI on a each day basis, most folks don't understand it. Whenever you talk to SIRI via an Apple device, you're using synthetic intelligence. Alexa from Amazon moreover makes use of the same technology. These devices, no matter the truth that lovely, are taken into consideration to be what is referred to as 'narrow AI' or a 'susceptible AI.' They are quality programmed to perform a very small range of obligations. These really represent the number one introductory steps of AI on an prolonged-term aim inside the course of a greater human interaction with machines. With that stated, apart from the fears that many human beings have about AI, there are numerous blessings that we're able to gain from those advantages.

Chapter 9: Neural Networks Inside The Future

It's hard to imagine the opportunities the future holds for neural networks. Because of the way this period is already integrating themselves into each issue of our lives, the potential for logo spanking new and revolutionary thoughts is higher than ever. We can envision a Jetson-like worldwide in which we will have self-using automobiles in preference to GPS gadgets that want to be programmed with our meant excursion spot. Imagine a car that has decided out your non-public alternatives inside the tune you be aware of, the temperature you're maximum cushty with, and a way to flawlessly alter your seat.

But all of that is feasible now. The future holds masses extra opportunities wherein neural networks can be carried out.

There are specific processes that this new studying technology can develop. One vicinity is within the discipline of digital intelligence. This form of utility might be deliberate, managed, predictable, and will in the end emerge as the following evolutionary step in artificial intelligence.

This sort of intelligence might be even towards matching its thinking and gaining knowledge of patterns to human beings. A device that could evolve and boom with mankind, adapt to the equal environment and studies from its reviews is inevitable.

To improve to date, however, requires era that could truely understand and make the essential changes to bridge the gap that now exists amongst AI and VI. As this new generation slowly engages in our global, extra of our sports activities can be completed out in virtual fact. We'll locate ourselves spending extra time with computer systems, giving the hundreds of data to percent. We'll communicate thru using avatars, social structures, and video video video games.

These virtual worlds will want to be created even though, but these are places wherein it is steady to study, strive, and fail at our attempts to decorate. They will take the vicinity of social structures and give the opportunity for us to hone our talents in business enterprise, finance, or even romance. Whatever you need to test out, there may be a virtual global to artwork in earlier than you're making your idea mainstream.

This will in the end end up a very automated international however no longer a self-conscious global as many human beings fear. Humans will however set the parameters and located limits on the type of things they need computer systems to do. Their smart software application software program can be capable of simplify and decorate our actual lifestyles however now not take rate and manage it. As lengthy as human beings located limits at the computer's capability to make bigger, the future will remain colourful for this technological improvement.

Right now, synthetic intelligence remains in its infancy, the subsequent decade may be a actual eye-opener. Not very a protracted manner in the future, we are able to begin to see the ones machines trade the way cars and planes are designed, how they will be operated, and how they may interact with humans. We will watch our days of exploration float similarly and similarly into area. In time, we are capable of witness the colonization of new worlds. This time literally.

The future moreover has many modifications in hold for a rustic's military might possibly. Soon, there gained't be a want for "boots at the floor" while a rustic is at warfare. One soldier may be capable of manipulate a whole fleet of drones an amazing manner to combat of their vicinity. These are already in use in some partial form now. Called unmanned aerial motors or UAVs those drones are able to being operated from a miles off vicinity and responding to a myriad of commands. In time, those UAVs will become self retaining and art work with out the useful resource of human path.

What does this imply? Imagine a fleet of drones all headed for a unmarried goal. If one drone is destroyed thru enemy fireside, the final drones must automatically reassemble and preserve at once to carry out their mission. Their capacity to have a look at and make bigger will permit them to comply to the characteristic of the destroyed drone and contain his assigned venture into their programming.

It is expected that with every new gadget brought, machine programming will growth in its complexity and abilities. Today, we suppose that artificial intelligence is one of the maximum fascinating forms of generation mentioned to guy. What are we able to assume while virtual intelligence turns into available to the mainstream population? These machines can be greater capable of interacting with humans and could revolutionize every issue of our each day lifestyles.

Another region wherein this new era will beautify human lifestyles is within the region of disaster response. Areas dangerous for people to go into can now be accessed thru deploying machines to carry useful aid to people who are lessen off from the rest of the area through way of catastrophic sports. Imagine how the ones realistic programs can be mounted in machines which could look for life underneath the rubble of ruined houses. How food sources may be brought quick and successfully. How rebuilding efforts is probably an lousy lot quicker and the manner the remedy of the injured might be executed fast and efficiently.

We'll see this period inside the movie industry, track, in agriculture, and in an limitless parade of different industries as time progresses. Right now, we're quite certain of what the destiny holds for a neural community and all of its many applications. What we aren't remarkable of is how short humankind will encompass it. No doubt, it will possibly be the more youthful and extra adventurous technology an amazing manner to encompass it first. They can be those to harness its exceptional functionality and they will be those who will have to set its limits.

Science has plenty to offer us in the manner of advanced pc era. The machines that allows you to be produced the following day and, in the years, to go back will open the door to a whole new international of adventure. But it will appear because of the truth humans are pushed by using manner of way of the powerful pressure of human desire to constantly locate better methods to do subjects are more potent than the numerous which might be effective and in time, they'll make the technology fiction of the past turn out to be the truth of in recent times.

Implementation and Interpretation

You've analyzed your facts, decided on an set of policies, and commenced education it with the type of facts you need to use it to interpret. Even after your gadget mastering algorithm is all professional up and prepared, even though, there are a pair extra property you want to do earlier than you're sincerely ready to start setting your smart software program program to paintings.

Analyzing the facts will can help you recognise what type of set of rules to use to interpret it, however in advance than you may truely use the set of regulations, you'll possibly need to perform a hint prep paintings to your information. Properly making geared up your information permits to make certain that you get the outcomes you're looking for and that the set of rules features the manner you suggest.

How an lousy lot education you'll want to do is predicated upon at the individual of the statistics you're running with. In the case of mainly large portions of unlabeled information, you could need to run an unmonitored set of guidelines on it first to emerge as higher familiar with the underlying form and help making a decision how satisfactory to utilize it.

The quantity and varieties of capabilities and attributes you want to bear in thoughts may additionally even have an impact on how hundreds education paintings you need to do for your records. If there are a number of lacking features or outliers, cleansing up the statistics can help your models run extra correctly. You may additionally want to transform the information by means of compiling it or scaling so it's less complex for this system to approach.